Worm

Animal

Series editor: Jonathan Burt

Worm

Kevin Butt

REAKTION BOOKS

Published by
REAKTION BOOKS LTD
Unit 32, Waterside
44–48 Wharf Road
London N1 7UX, UK
www.reaktionbooks.co.uk

First published 2023
Copyright © Kevin Butt 2023

Printed and bound in India by Replika Press Pvt. Ltd

A catalogue record for this book is available from the British Library

ISBN 978 1 78914 794 0

Contents

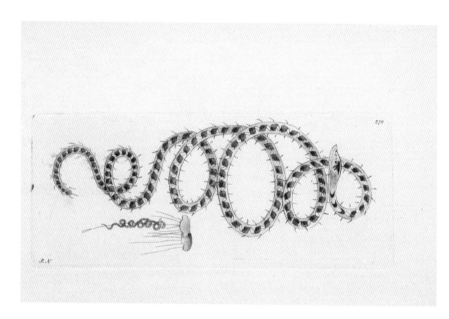

1 Introducing the Worm

When collecting earthworms, usually a little digging is required. Alternatively a liquid is used to drive some of the deeper soil dwellers out of their burrows. The worms are then placed in a container. Something very different, however, may be needed for some of the larger, more robust species. Standard earthworm collection techniques have proven to be of no use when collecting *Lumbricus badensis*, a species endemic to the Black Forest in Germany.[1] Personal experience, based on advice from local retired researchers, showed that the only way to capture adults was with stealth and strong will.[2] At night, under red light illumination, it is possible to observe the first 30 centimetres (1 ft) of these worms extending from burrows, feeding at the soil surface. With soft tread and a rapid two-handed 'grab and hold' technique, one can prevent the worm from retreating into its burrow. Thereafter it becomes a delicate tug of war, although tugging is only possible when the worm has relaxed some muscles in its attempt to retreat. Over five to ten minutes the worm can slowly be extracted to reveal an adult with a biomass of up to 40 grams (1 1/2 oz) – some ten times heavier than the better known *Lumbricus terrestris* – and up to 60 centimetres (2 ft) in length. Some earthworms still surprise me, even after researching this group for more than thirty years.

Over time many organisms have been given the label 'worm'. Historically, this stems from a time when scientists, including

Illustration of the *Ophidonais serpentina* worm, signed RN (Richard Nodder). Plate engraving from George Shaw and Frederick Nodder, *The Naturalist's Miscellany* (1796).

Linnaeus, responsible for the binomial system of naming living organisms that we still use today, had begun to try and classify all living animals.[3] Most were easily assigned to the more obvious groups such as mammals (fur-bearing and suckling), birds (feathered and egg-laying) and fish (scaled and with gills), and still recognized after three centuries, but some proved less than straightforward. One group, *Vermes*, brought together many things that didn't obviously fit elsewhere and may have led to the evolution of the term 'worm', which is now in everyday usage. It is a name that appears and is recognizable in many languages – *wurm, orm, wjirm, wuerm, vierme* – to provide just a few. All of these describe a cylindrical animal with no limbs and few obvious external features, apart from a mouth and an anus. In general terms, similarities can be seen, which may be a function of convergent evolution, but from a biological standpoint these animals may differ greatly, having evolved many millions of years apart, and belong to a dozen different invertebrate phyla and other groups.

A collection (a squirm) of green earthworms (*Allolobophora chlorotica*), a common British species.

Before considering specific organisms that are referred to as worms, however, a deeper conceptual understanding of 'worm' may be appropriate. Humans and worms have an intimate and closely entwined history that is both physical and cultural. In various guises worms have been portrayed within a spectrum ranging from the benign, and perhaps even beautiful, to evil, cankerous and deathly. The following explores some of this range.

There is a history of 'worm science'. Probably even before investigations by ancient Greeks and Romans, it was understood that there were organisms living in our bodies. Many of these were referred to as, and indeed were, worms. Daniel Le Clerc in the early eighteenth century brought together clinical observations from across the ages and interpreted what was then known.[4] Fundamentally, three worm types had been determined and were referred to as Teretes (common round worms), Ascarides (small white worms) and Lati (tapeworms or flatworms). The last were also known as gourd worms by the Arabs as they were similar in shape to such seeds. It was thought from Roman times that these worms bred in the guts of humans, and Pliny the Elder knew that they reached 9 metres (30 ft) in length. However, there was no awareness of the life cycles of human parasitic worms until many centuries later. It was not determined until 1668 that people who ate 'measly beef' would become infected with tapeworms.[5] Such larval tapeworm stages (cysts) were known by Aristotle in the fourth century BC to be present in 'measly pork' but were not recognized as a critical part of the tapeworm life cycle. It is no wonder that we have a fascination with parasitic worms, as in the short history of human evolution, we have become host to three hundred different species, including more than forty types of tapeworm.[6] Of these parasites, *Ascaris lumbricoides*, the large roundworm, is common and one of the six worms listed and named by Linnaeus, and whose scientific name has been unchanged since.

Fourteenth-century depiction of worms (*vermis*) from Jacob van Maerlant, *Der naturen bloeme* (c. 1350). It is difficult to be certain, but perhaps the worms are shown emerging from their burrows in the soil?

One of Britain's great scientists, Robert Hooke (1635–1703), was able to view details of small organisms with his self-built microscope, which prompted him to speculate on the role of one of these 'worms' in 1665:

> As among greater Animals there are many that are scaled, both for ornament and defence, so are there not wanting such among the lesser bodies of Insects, whereof this little creature gives us an Instance. It is a small white Silver-shining Worm or Moth, which I found much conversant among Books and Papers, and thus suppos'd to be that which corrodes and eats holes through the leaves and covers; it appears to the naked eye, a glistening Pearl-colour'd Moth, which upon the removing of Books and Papers in the Summer, is often observ'd very nimbly to scud, and pack away to some lurking crannery, where it may the better protect it self from any appearing dangers.[7]

From Hooke's description and associated excellent drawing we would now recognize this organism as a silverfish, a wingless

Spined marine worms, from Ernst Haeckel, *Art Forms in Nature* (1904).

Robert Hooke's 'small white silver-shining worm or Moth', from *Micrographia* (1665). This wingless insect is now commonly known as a silverfish.

insect, which lives in damp conditions and may still be found in bathrooms. Its segmented and apterygote appearance led to the nominal worm tag.

A Fellow of the Royal Society, Antonie van Leeuwenhoek (1632–1723), was another pioneer of microscopy who 'thought fit to consider of the Animals that our Butchers call Maggots, which often are found in the Livers of Sheep, when we have wet summers.' He further wrote that:

> I did fetch out of a gall vessel one of these animals, which was twice as long in the vein, and was indifferent small. When I had discovered this, I did fancy that the worm had forced herself into the small vessel, so that it could not get back again and therefore did dye, which had first caused a corruption in that part; and secondly, the vein being stopt up by the worm did cause a second putrifaction and matter.[8]

These worms are what we now refer to as liver flukes and are common parasites of sheep.

Moreover, worms are more than the free living or parasitic entities that permeate our lives. They have entered our psyche and found their way into thinking and writing over the centuries. The ideas worms conjure tend to fall into categories that relate to illness, death, decay, revitalization and sexual association, although the boundaries between these categories can often become blurred.

During the Romantic movement of the late eighteenth and early nineteenth centuries, the beauty of nature was celebrated, medievalism was revived, and science treated with some suspicion. Numerous literary notables, including William Blake (1757–1827) and Lord Byron (1788–1824), produced works that featured worms, but here the worms were not necessarily viewed as we might appreciate them today. They were described as serpents, potential agents of death or even sirens drawing sexual pleasure from unsuspecting actors:

And, oh! That pang where more than Madness lies –
The Worm that will not sleep – and never dies –
Thought of the gloomy day and ghastly night,
That dreads the darkness, and yet loathes the light –
That winds around, and tears the quiv'ring heart –
Ah! Wherefore not consume it – and depart![9]

This omen, from Byron's 'The Bride of Abydos', foresees the unfortunate outcome of a forbidden love match that ends in the death of both potential partners. Disaster is foretold. 'The Worm that will not sleep' could be interpreted as the eating away of something (love) that was never likely to be consummated.

Perhaps a more direct reference to – and one of the most striking images of – a worm-based creature is the 'invisible worm' in William Blake's 'The Sick Rose', which alludes to an animal that, if taken at face value, could be perceived as a caterpillar, one which brings about the decay and ultimate demise of a once colourful flower:

O Rose thou art sick.
The invisible worm,
That flies in the night
In the howling storm:

Has found out thy bed
Of crimson joy:
And his dark secret love
Does thy life Destroy.[10]

The symbolism within this enigmatic poem, however, is deeper than this interpretation, as the worm could represent the destruction of a relationship and the corruption of something beautiful. The bed could be taken simply as a flower bed, but perhaps in the romantic sense it may be that used by two lovers with reference to an initial meeting and sexual union. Here, the worm may depict the deadly side of an infection or disease and degradation and at the same time could be linked to a suggestion of carnal pleasure, something preying on the innocence of the natural bloom.

Death stalks us all and is inevitable. A recurring theme in art that peaked around 1500 was the memento mori, macabre imagery that reminded viewers of death, often through representations of a skull and associated actors, including worms. Sometimes depicted as snakes, these worms would sit as a reminder that in death our physical bodies decay and the very elements from which they are made are consumed by maggots and worms to give rise to new life: organic recycling given an artistic and perhaps even spiritual take. Exquisite artworks, in the form of bejewelled skulls, carved ivory and paintings, act as a reminder of human mortality and the brevity of life. In addition, reference to our mortality was made clear in the mid-fifteenth-century poem 'A Disputation between the Body and the Worms', in which the narrator recounts a debate, experienced within a dream, between a once beautiful but now dead woman and the worms that are consuming her:

> We only ask your flesh on which to feed.
> For we have no way of tasting or smelling
> Your horrible, rotting stinking waste.
> All creatures find you extremely repelling
> Except for us worms; we're already disgraced.[11]

It is interesting to hear the collective voice of the worms arguing their case to eat the woman's decaying body. Worms have consumed our 'rotting' flesh throughout time immemorial, feeding on notable figures and insignificant people alike. What is natural to worms is distasteful and abject to us. It is thought that the first reference to decaying bodies as 'food for worms' comes from the *Ancrene Riwle* of circa 1220, a monastic treatise in which an unknown author, writing advice for three sisters who have retired to a life of prayer and penance, states, 'Ne schalt tu beon wurmes fode?' Shakespeare appears to have taken and developed this idea when, in *Henry IV, Part 1*, the mortally wounded Hotspur starts to address himself with the words 'No, Percy, thou art dust, and food for . . .', before dying. Prince Hal completes the sentence, 'For worms, brave Percy' (v.4).

Many others have furthered this thinking by suggesting that worms are not simply associated with death, but embody death and are metaphors for finality. An example is found in the final verse of a poem by Edgar Allan Poe:

Memento mori by unknown artist, c. 146–95, engraving.

Out – out are the lights – out all!
And, over each quivering form,
The curtain, a funeral pall,
Comes down with the rush of a storm,
While the angels, all pallid and wan,
Uprising, unveiling, affirm
That the play is the tragedy, 'Man,'
And its hero, the Conqueror Worm.[12]

Here, using the theatre as a backdrop for life, humans are depicted as actors incapable of preventing their inevitable demise, no matter what benevolent extraordinary or supernatural forces are around them.

Worms are conquerors that exploit devastation and catastrophe. Humans and other creatures struggle continuously for survival, yet we inevitably in death return to the ground, the worms' abode, where our lifeless bodies become the very sustenance of their small but vital lives.

Worms feature in science and all forms of writing and other arts. Before considering these representations though, we will look at some of the creatures within the natural world that are currently referred to as worms. It took until the late nineteenth century before 'Vermes' were finally disentangled and 'A Natural History of Worms' was finally brought together with subdivisions recognizing the heterogeneity of this diverse collection of animals.[13] Here, though, we will take a more general approach and consider any creature commonly called a worm.

'Worm' is applied to many insect larvae, to parasites (round, flat and thread worms, as we now know them) and those that dwell in subterranean substrates including soil and sand. These include nematodes, pot worms and annelids, which are the true segmented worms, comprising marine polychaetes and terrestrial oligochaetes. The latter (earthworms) will form the focus of this book, but in the true spirit of diversity it is first worth exploring some of the other worms that are found worldwide. Some of these are reasonably well known, but a few may be surprising, for example, the worm tag also extends to some vertebrates, such as worm-like amphibians and reptiles. Most of the examples of worms given, which are only a fraction of those in existence, are closely associated with humans and therefore of direct interest.

Most people will be familiar with nematode worms (often called roundworms) as many are free-living and found within the soil. Nevertheless, the majority of nematodes tend to be very small, 0.1–1 millimetres (0.004–0.04 in.) in length, placing them in the microfauna group that include protozoans, such as amoeba. Nematodes account for the most numerous multicellular organisms found in ecosystems, with millions present in a single square metre of soil. Morphologically, they have a simple, cylindrical body shape that tapers at both ends. Their feeding apparatus has evolved for the consumption of bacteria, fungi, plant material or other animals, and the group has been subdivided by their type of feeding. In this way these worms have many trophic relationships with other organisms. Nematodes may seem too small for us to be concerned about, but they can cause numerous problems. Herbivorous species, such as eel worms and including potato cyst nematodes, are soil pests that directly attack and cause damage to plant roots in agricultural and silvicultural systems, and so need to be controlled. Some carnivorous nematode species, however, such as *Phasmarhabditis hermaphrodita*, are now used in horticulture as a form of biological control for slugs, so removing the need for more traditional and toxic slug pellets. The tiny nematodes enter the body of the slug, which consequently stops feeding. As the worms multiply within the slug, it crawls underground and dies. After feeding, the nematodes seek out other slugs on which to continue feasting. Developed as a commercial product by researchers from Bristol in the 1990s, 'Nemaslug' is now available to gardeners as a sachet of moist clay containing millions of the worms. Users simply mix the clay with water and apply it to the soil via a watering can. These nematodes may also kill snails but are not harmful to other beneficial soil organisms or animals higher in the food chain. Recent research has shown that this form of biological control is as effective as more traditional measures.[14]

One nematode worm of particular importance is *Caenorhabditis elegans,* a non-hazardous, non-pathogenic, non-parasitic species that lives in temperate soil environments. At 1 millimetre long, this free-living, transparent animal may seem insignificant, but is noteworthy as it was the first multicellular organism to have its whole genome of DNA sequenced, in 1988, and its connectome (neural wiring) determined, in 2019. *C. elegans* has therefore become a model organism that is used extensively in molecular and developmental biology. Small does not mean unimportant. A whole website is available to view current developments in research associated with this species, including a database of genetics, genomics and phenotypic information that is constantly updated.[15] This species also holds another claim to fame: representatives of *C. elegans* visited the International Space Station from 5 December 2018 to 13 January 2019 as part of an experiment to understand why astronauts lose some of their muscle mass in space. Since these miniature worms are sufficiently like humans to enable comparison of essential biological characteristics, *C. elegans* helped to determine the effects of changes brought about in space, including alterations to muscle and the ability to use energy.[16]

Other nematodes may have a more direct bearing on our lives, as, for example, parasitic worms (helminths), which are able to live within vertebrate hosts. Unlike soil-dwelling nematodes, parasitic worms can be much larger: *Ascaris lumbricoides*, for example, can reach 30 centimetres (1 ft) in length and live in large numbers within the human intestines, where it derives its nutrition from the surrounding food in which it is bathed. This is the largest and most common parasitic worm in humans and may affect more than 1 billion people worldwide.[17] These worms are prevalent in unsanitary sites and commonly spread in places where untreated human faeces are deposited directly on to agricultural land. Treatment with appropriate drugs can be simple and efficacious, but effective sanitation is required to prevent reinfection.

Threadworms, also known as pinworms, are another quite common helminth. These are smaller white worms, about 1 millimetre in length and pointed at one end (like a pin). Typified by *Enterobius vermicularis*, these very common human endoparasites are found worldwide in both developed and developing countries. Further parasitic nematodes include hookworms, such as *Necator*

Worms in space: original logo of the European Space Agency's molecular muscle experiment using the nematode *Caenorhabditis elegans* to assess effects of weightlessness on worms and indeed humans.

Ascaris lumbricoides nematode worms egested by a single infected child following treatment. This parasite inhabits the human small intestine and is spread from human to human by the faecal–oral route.

americanus and *Ancylostoma duodenale*. These intestinal nematodes produce a wide range of symptoms including diarrhoea, abdominal pain and general weakness. Hookworms specifically cause chronic intestinal blood loss that ultimately results in anaemia. The World Health Organization (WHO) estimates that more than 880 million children annually need treatment for ailments caused by these parasites.[18] As with *Ascaris*, these parasitic worms and the problems they cause can be easily treated, if appropriate anthelminthic pharmaceuticals are available.

Flatworms (Platyhelminthes) represent another invertebrate taxon, members of which also utilize humans as hosts for their parasitic lifestyle. Tapeworms, for example, may be well known to most of us from biology classes at school, general reading or via unfortunate personal experience. These relatively simple, highly segmented organisms are flattened in shape to allow oxygen and nutrients to diffuse directly into their bodies as they have no specialized respiratory or circulatory systems. The scolex (head) of the worm hooks into the lining of the gut and the body grows and exists directly within the alimentary canal. The beef tapeworm (*Taenia saginata*) is a prime example of this class of worm;

it infects humans, who act as the definitive host, by first infecting cattle, which are the intermediate host and support the tapeworm's larval stages. Infection of humans can only occur if undercooked beef is eaten and the larvae is able to reach the gut. Infection with *T. saginata* may cause gastrointestinal upset, but a well-fed person might not know that they are host to such a parasite, the only evidence being a few segments of the parasite in their faeces – however, when the worm dies, probably after several years, the scolex detaches and the worm is egested in full. An adult beef tapeworm is normally 4–10 metres (13–32 ft) long, but can reach 22 metres (72 ft) in length. The 'passing' of such a worm, in more than one sense of the word, would not go unnoticed.

Another type of flatworm that has caused some concern in Britain since the 1960s, but perhaps less directly, is the accidentally introduced New Zealand flatworm *Arthurdendyus triangulatus*. This 10-centimetre-long (4 in.) soil-dwelling predator probably arrived in the egg stage of its life cycle with imported plants. This exotic species is a problem because it preys on earthworms and has decimated their populations in some areas of Northern Ireland and Scotland.[19] With no natural predators of its own in Britain, soils where this species are found can therefore suffer due to the absence of earthworms. This worm is often found under debris on the soil surface in gardens and at the margins of agriculture. To feed, it exudes enzymes from its skin when beside an earthworm and then ingests the externally pre-digested food. It appears that little effective control of these invaders can be found, but environmental conditions, mainly driven by soil moisture content, may restrict the spread of this non-native species to the whole of Britain. In areas where the flatworm has become established it may enter classical predator–prey population oscillations, where predator numbers are influenced and potentially restricted by prey numbers. Without food it might be thought that the flatworm problem would disappear, but one advantage in such interactions is that the flatworm can sit out times when no prey is around by entering periods of de-growth. It does this by reabsorbing parts of its own body, such as the reproductive organs, that are not required for survival. Once food (in the form of earthworms) becomes available again, the New Zealand flatworm can feast, regrow absorbed organs and exist as before. Having a relatively simple body plan and strategies that have evolved for periods of resource shortage can be extremely advantageous to some worms.

In the soil there are small unpigmented worms (typically 10–20 millimetres ($2/5$–$4/5$ in.) long) often referred to as pot worms, but more formally as the Enchytraeidae. These may occur in large numbers (400–800 per square metre) and are often found in acidic soil conditions (pH less than 7) that do not suit their larger earthworm 'cousins'. Anatomically they are quite similar to earthworms, but in miniature, with a transparent body wall. Pot worms eat small mineral and organic fractions of the soil

An adult beef tapeworm with prominent head (scolex). In a well-fed human, the presence of such a 'passenger' could go unnoticed for most of the worm's life.

Australian flatworm (*Australoplana sanguinea*). This small invertebrate, another predator of earthworms, is now present in Britain. The effects of its introduction to Britain is under investigation.

New Zealand flatworm (*Arthurdendyus triangulatus*) on the author's hand. This invasive alien species in Britain is a predator of earthworms, whose populations it may significantly affect.

– their diet is probably based mainly on the microbiota (bacteria) present on the organic fragments. These worms go unnoticed by most, except keen gardeners who may see them when turning over organic-rich soils. Their faecal pellets may be influential in soils at a micro-scale, but this group is under-researched and little is known of their burrowing, for example. Where found, they may occur in large numbers due to clustering in favourable spots.

Some other worms that are rarely seen but leave obvious signs of activity are related to earthworms, but marine in nature. Polychaete worms live in sandy substrates. Small coils of piled material that can be observed in the intertidal zone at low tide are the castings of lugworms (*Arenicola marina*), which live in U-shaped burrows. Look closely and you will also see a depression in the sand about 8 centimetres (3 in.) away from the casting, which shows the second end of the burrow, from where the lugworm sucks in sand to extract nutrients, before egesting the castings out of the burrow at the other end. Lugworms are collected by sea fishermen as bait and legally dug from the shore in large numbers for personal use. Nevertheless, more recently in southern England, this practice has been subject to scrutiny by an Inshore Fisheries and Conservation Authority (IFCA) that is working to prevent commercial bait digging, which was found to be decimating stocks in some areas.[20] It is good to learn that legal measures can be used to protect even the lowliest of animals, each of which, we must remember, plays a critical role in its ecosystem.

A worm native to Eurasia and only seen from March to October is the slow worm. This shiny animal, golden-silver grey in colour, grows to 50 centimetres (20 in.) in length. Unlike the worms considered so far, the slow worm is a vertebrate and a member of the class Reptilia. *Anguis fragilis*, also known as the blindworm, is a harmless, legless lizard with a passing resemblance to a snake. However, this slow worm is neither blind nor slow and possesses functional eyes and eyelids that can blink, clearly showing that it is not a snake. Hibernating in winter under leaf piles or among tree roots, slow worms prefer to live in heathland, tussocky grassland and woodland edges where they predate snails, insects, spiders and earthworms. Slow worms can live for twenty years and like some other lizards incubate their eggs internally and produce live young in late summer. They are rarely found in gardens as they can fall prey to domestic cats, but if attacked by a predator they can, as other lizards, shed their tail in an attempt to confuse the predator, escape and then re-grow a new tail.

A further vertebrate group that is even more worm-like comprises the lesser-known caecilians, found in humid tropical regions, mainly of South and Central America, Africa and southern Asia. These rarely seen, worm-like amphibians form the order Apoda (no legs), alongside the better studied frogs, toads and salamanders. Most caecilians are fossorial (live underground), are found in leaf litter or frequent moist soils, usually close to streams and lakes. Depending on species, adults range from 10 to 150 centimetres (4–60 in.) in length and their elongate bodies, with distinct segmentation, superficially resemble large earthworms. Their eyes are reduced and covered with skin and the skull is heavily ossified, so adapted to a burrowing lifestyle. These are the only vertebrates known to use their entire body as a

Casting of the marine lugworm (*Arenicola marina*) revealed in sand at low tide. The depression is the feeding end of the U-shaped burrow in which the worm lives.

A slow worm
(*Anguis fragilis*),
also known as a
blindworm. This
western Eurasian
reptile is neither
slow nor blind,
nor even a worm.

hydrostatic system for locomotion (like invertebrates such as earthworms). Caecilians prey on soil-dwelling insects and on earthworms, using needle-sharp teeth to hold the prey before it is swallowed whole. Some of the smallest caecilians, such as *Idiocranium russeli*, were initially mistaken for earthworms because of the close morphological resemblance brought about through convergent evolution to a burrowing lifestyle. Only microscopic observations and dissection revealed the true nature of their classification. Due to their small size and relatively unexplored habitat, new species, such as *Caecilia pulchraserrana*, are still being discovered.[21]

A return to the invertebrate world shows that there are numerous worms that are the larval forms of insects. Prior to complete metamorphosis, the life stage that hatches from the egg may

A very worm-like caecilian (*Microcaecilia dermatophaga*), a small, legless amphibian found in tropical wetland soils.

Close-up of *Caecilia pulchraserrana* that shows reduced external features evolved for a burrowing existence. These animals are known locally as blind snakes or captain worms.

appear worm-like, with less than obvious insect legs and certainly no wings. One such creature, comprising many species, which has been described in song as 'measuring the marigolds', is the inchworm. This is the larval form of a family of moths called the Geometridae, a name derived from the Greek for earth (*geo*) and measure (*metron*), as the caterpillars, also known as loopers, appear to measure the earth in inch-like steps as they move along. The characteristic looping gait is achieved by extending the anterior part of the body and then bringing the rear up to meet it, so inching along. These forest caterpillars feed on leaves, particularly of trees, and can become serious pests. They are often predated by birds but can camouflage themselves by remaining still and thereby resembling leaf stalks or small twigs. The inchworm stage of the moth's life ends when it pupates to form a chrysalis and eventually emerges as an adult insect.

The silkworm, another well-known insect, is the caterpillar of a silk moth (*Bombyx mori*), a species that no longer lives in the wild after domestication in China many thousands of years ago. The silkworm is of commercial interest as the single thread of raw silk that it weaves around itself during pupation is the basis of a

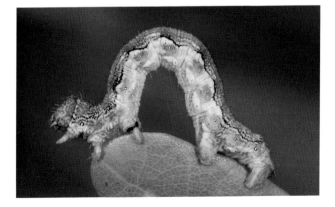

Caterpillar of the mottled umber moth (*Erannis defoliaria*), which is also known as either a looper caterpillar or an inchworm, due to its mode of locomotion.

Silkworm (*Bombyx mori*) caterpillars feeding on mulberry leaves. These moths were domesticated by humans for silk production thousands of years ago.

multimillion-dollar industry. Each thread can be up to 900 metres (3,000 ft) in length and is obtained by carefully unravelling the pupa after first killing the worm with steam. Nearly 5,000 pupal threads are required to produce 450 grams (16 oz) of silk. The silkworm larvae are raised by feeding them on mulberry leaves. As a valuable product, silk was historically traded along the Silk Road that reached from China across to Europe; China still acts globally as the principal producer of silk. A recent innovation has been to modify silkworms genetically with spider genes to improve the strength and flexibility of the silk produced.[22] Silkworms are also sometimes kept as pets, particularly for children, as they are easy to feed – they also allow children the chance to witness the life cycle of a moth at first hand.

The glow-worm (family Lampyridae) is an insect with which people may be familiar from first-hand observation in nature. Here though, it is only the wingless adult females that carry this common name and produce a light to attract the flying males. The name 'glow-worm' is something of a misnomer as the insect is, in fact, a beetle, about 3 centimetres (1 in.) long, but the adult female does actually have a passing resemblance to a worm, as her abdominal segmentation is more apparent than the male's

A female glow-worm (*Lampyris noctiluca*) using bioluminescence to attract a mate. The clearly seen segmentation of the wingless female gives a worm-like appearance.

without a covering of wings. The full life cycle of the glow-worm may extend over two to three years, but the majority of this is spent as a reclusive larva that feeds on small snails in the undergrowth and in the soil. The light from an adult female is seen during a period of just a few weeks after emergence from a chrysalis, when she does not feed and has only one objective, to mate and produce the next glow-worm generation. The cold light is a form of bioluminescence and produced by enzyme-catalysed biochemical reactions within the abdomen of the female.

Another insect larva, which was once more widespread and associated with humans, is woodworm. Evidence of this species, also known as the common furniture beetle or *Anobium punctatum*, can be seen in old furniture and wooden fittings where circular exit holes 1.5–2 millimetres in diameter show that adults have emerged. The larvae live within and eat wood over a three- to five-year period, whereas the non-feeding adults emerge, mate, lay eggs and die within only two to four weeks. It is not so common these days to hear mention of furniture as 'having worm', due to chemical prevention treatments of wood, but I recall asking, as a child, how and why these small perfectly circular holes had been made in an old wooden box. Even in a marine setting, wood is not safe from attack by 'worms', though once again this is a case of misleading naming. More of a problem when ships were constructed primarily of timber, ship-worm (for example, *Teredo navalis*) is a form of bivalve mollusc that can attach itself to the hull of wooden vessels and, if untreated, can lead to the destruction and sinking of ships as the worms burrow inwards. During the eighteenth century this was one reason why the hulls of many wooden ships were clad with a copper covering. Similarly, wooden piers and docks in a marine setting can also suffer from such attack and need protection.

A little known but evolutionarily intriguing animal is the velvet worm (*Peripatus*). This 2–15-centimetre-long invertebrate with numerous stumpy legs is found in the tropics and, significantly, is thought to represent a link between annelids (segmented worms) and modern arthropods (insects, arachnids, crustaceans and myriapods). With large antennae, the velvet worm is a

Circular holes in a wooden artefact caused by woodworm (the larval forms of wood-boring beetles (*Anobium punctatum*), also called furniture beetles).

nocturnal predator that hunts by touch for insects, such as crickets. Remarkably, when potential prey is encountered, the velvet worm captures it by squirting a glue-like protein from two specialized nozzles close to its mouth. The immobilized prey is then fed upon at leisure and in safety. Not only does *Peripatus* eat the prey, it also eats the glue and recycles it for reuse. These worm-like animals may have changed very little in the past 570 million years.[23]

In the eyes of many, size really matters. When thinking of worms, the parasitic flatworms are lengthy, but what of free-living worms? Temperate regions have relatively little to offer in this regard when it comes to earthworms, but venture to a region of Victoria in southeastern Australia and it is possible to see (or perhaps hear) a rare worm, listed by the International Union for the Conservation of Nature (IUCN) as 'vulnerable'. This is the giant

New Zealand peripatus (*Peripatoides novaezealandiae*), a type of velvet worm that may represent an evolutionary link between annelids and arthropods.

Gippsland earthworm (*Megascolides australis*). Reliably measured at an average length of 75 centimetres (30 in.) when partially contracted after excavation, this species can extend to 1–2 metres (3–6 ft) and weigh up to 380 grams (13 oz). Rarely extracted from the soil, due to its large size, *M. australis* can be detected by gurgling sounds that the worms sometimes make when retreating down their wet tunnels. These tunnels, like the worms themselves, are approximately 2 centimetres (4/5 in.) in diameter and can be found running horizontally within 10 centimetres (4 in.) of the soil surface in the heavy clays close to creeks. However, they tend to be located deeper in summer, as soils dry. These worms are little studied, and their long-term conservation status is uncertain due to changes in agricultural practices and soil management.[24]

Perhaps the longest worm ever recorded is the ribbon worm (*Lineus longissimus*). Also known as the bootlace worm, this unsegmented aquatic invertebrate is a member of the Nemertea. Most species in this group are marine and less than 20 centimetres (8 in.) long, but *L. longissimus* can reach 30 metres (100 ft) in length. Nemerteans are the simplest animals that have a circulatory system and a gut with a mouth and an anus. These carnivorous animals possess a needle-like stylet which can be everted to assist with prey capture. The body of the ribbon worm is darkly pigmented and may be given a purple iridescence by a covering of small hairs (cilia), with eye spots (to detect light) present on a rectangular head. In Britain this species can be seen around the coast, often located in intertidal rock pools entangled around kelp. All ribbon worms can reproduce sexually, may regenerate lost or damaged body parts and some species can break their bodies to form fragments, which then grow into separate, complete individuals. Little research has been undertaken on this large Nemertean, but others have been more closely studied, with DNA evidence uncovering a decades-old mystery.

In Lancashire, a county in northwest England, there was thought to be a single endemic organism, 'the Croston Worm', a Nemertean known from one small lake in Croston, near Preston. This had been identified from histological (tissue) examination in the 1970s and named *Prostoma jenningsi*, a new species, and later listed in the British Red Data Book as 'Insufficiently Known', as well as a 'Species of Principal Importance' under the 2006 UK Natural Environment and Rural Communities Act. Investigations between 2010 and 2020 found Nemerteans in this location and nearby ponds. DNA sequencing of nuclear 18s ribosomal RNA and mitochondrial Cytochrome Oxidase I (COI) genes was undertaken, and phylogenetic analyses performed to establish the taxonomic status of recovered specimens. All available *Prostoma* sequences (*Prostoma eilhardi* and *Prostoma graecense*) were downloaded from GenBank® and Barcode of Life Data System (BOLD) databases for comparison. Results showed that the *Prostoma* recovered from the Croston site and all other locations in Lancashire were not distinct from *P. eilhardi* and *P. graecense*, which suggested there was a strong case for the species status of *P. jenningsi* to be revoked.[25] A case of good science from improved investigative techniques but resulting in a change in the status of the only species, and a worm at that, once thought to have been endemic to Lancashire.

So far we have looked at something of what 'worm' can refer to in terms of historical thinking and modern science. Nevertheless, if questioned about worms, we might first think of earthworms. This is for reasons to be explored in Chapter Two. Many people are not fully aware of how important earthworms are in the formation, maintenance and functioning of soils on this planet. The renowned scientist Charles Darwin stated that: 'It may be doubted whether there are many other animals which have played so important a part in the history of the world, as have these lowly

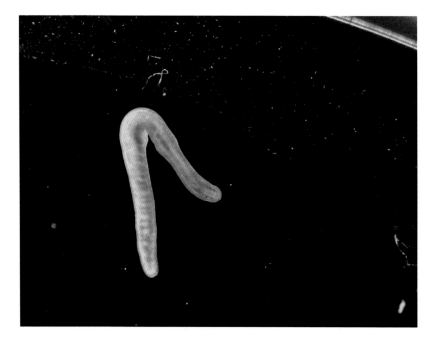

organised creatures.'[26] The reasons why Darwin wrote this will be detailed later, but they provide justification for why this book looks specifically at earthworms rather than the other types of worm that have been mentioned. Some of these have merit in their contribution to life on Earth, but earthworms cannot be underestimated, as without their continued presence, many ecosystems would collapse.

The 'Croston worm': a tiny freshwater nemertean whose identity was finally determined through DNA analysis.

We have already encountered large examples of earthworms from the Black Forest and from Australia, but perhaps there is a need to explore some basics before embarking on a discourse relating to more detailed aspects of their lives, and how they have a bearing on our own. Earthworms are generally cylindrical with a pointed anterior end but a posterior that tends to be more rounded. They are members of the annelid group of invertebrates, so have segments (annuli), often giving their bodies a ridged appearance. The body of an adult earthworm has a clitellum (saddle or collar), positioned towards the anterior or as much as halfway along the body: although also cylindrical, it is usually raised away from the rest of the body and covers several segments. Immature earthworms do not possess a saddle. The earthworm body comprises two sets of muscles that are circular and longitudinal within the earthworm and act antagonistically and as a part of the animal's hydrostatic skeleton, which works against the coelom, a body cavity. By peristaltic movement of these muscles, earthworms can move their bodies forwards or backwards and are assisted by hair-like setae (bristles) projecting from each segment that provide traction against the soil, whether in a burrow or on the soil surface.

Internally, earthworms possess a gut that runs from mouth to anus, subdivided into sections that allow for breakdown of their food and extraction of nutrients. A fact that is perhaps little known is that some earthworm species are able to collect materials by grasping them by their mouth, even with no teeth, and when foraging at the soil surface can move objects as heavy as their own body weight. Most of the vital organs in the body are present in

Anatomical earthworm model, commercially produced for use in education.

the anterior, with the rear of the earthworm primarily an extension of the gut. Earthworms possess a closed circulatory system with a major blood vessel present on the dorsal side. Some sources suggest that earthworms have multiple hearts (five or more), but these are simply swellings of vessels that circle the gut and assist with circulation of blood. Unlike many other organisms, earthworms do not have a system of gills or lungs to obtain oxygen for respiration. In the soil environment they can obtain required oxygen from the air by diffusion (passage against a concentration gradient) through their moist skin, which is then taken up by a network of blood capillaries. The skin is kept constantly moist by the secretion of mucus, making earthworms feel slimy to the touch. If mishandled, earthworms will produce more mucus and perhaps even exude a coelomic fluid through dorsal pores that can smell unpleasant and may act as a defence against some predators.

Earthworms possess a nervous system that is more highly developed towards the anterior. The first few segments are particularly well endowed with sensory cells, meaning that earthworms can respond to touch, a change of temperature, salt concentrations and acidity. Earthworms have been shown to be selective in their feeding and must therefore be able to 'taste' potential food materials. Their senses of smell and hearing are thought to be poor. Nevertheless, they will react rapidly to vibrations, perhaps not surprising for a soil-dwelling organism. Even though earthworms do not have eyes, they do possess light-sensitive cells, so can perceive the difference between darkness and light.

There is no such thing as a typical earthworm as they come in many sizes and colours. These may range from a few millimetres (for example, *Dendrobaena pygmaea*) to 2 metres (6½ ft) in length (*M. australis*). Colours tend to be relatively dull and within a range of pinks, reds and greys. However, even in temperate regions, green worms are not uncommon, such as *Allolobophora chlorotica*, and

A 'fried egg'
earthworm
(*Archipheretima
middletoni*) in
the leaf litter of
a forest in the
Philippines. The
colouration may
warn potential
predators of the
noxious nature
of this species.

the overall colour of the steel blue worm (*Octolasion cyaneum*) is contrasted by a bright orange clitellum and yellow tip to its tail. Striped (alternating coloured segments) are also relatively common in some species, such as *Eisenia fetida*. However, some of the more recent earthworm discoveries in tropical locations are perhaps the most striking. The so-called 'fried egg worm' (*Archipheretima middletoni*) from the Philippines has colouration that gives away its common name.[27] This 30-centimetre-long (12 in.) species, not discovered and described until 2009, does not burrow but lives in the leaf litter of the highland forests. It is thought that the patterns on the body may deter predators from eating this earthworm as it contains a toxin.

All earthworms are hermaphrodite (have both male and female reproductive organs), so can produce sperm and eggs. Nevertheless, the majority still engage in sexual reproduction. To achieve this, two adults are required, and they normally encounter each other below the soil surface. By unknown (possibly biochemical) means, they must determine that they are in the company of the same species and then set about mating. This involves one animal having the role of a male to the others female and vice versa. After they have aligned appropriately, with each anterior facing in opposite directions and overlapping by the required amount, they lock together. Each then produces sperm, which is rhythmically squeezed along a (seminal) groove on the outside of the animal but against its partner until it is collected in a sperm store. Once this is completed the two partners separate and go their own ways. To produce the next generation of earthworms, each partner then needs to make a cocoon. This is where the clitellum comes into play and works as a secretory organ producing a proteinaceous covering in the shape of a cylinder around itself. This short length of material is then shed underground by the worm as it withdraws backwards. As this occurs, the worm deposits its own

*Octolasion
cyaneum* (the
steel-blue worm),
a parthenogenetic
earthworm often
seen above ground
after rain.

eggs and nutritive material inside the cylinder with some of the collected sperm from the mating. As the head of the worm finally withdraws, the two ends of the cylinder close to form a spherical or ovoid cocoon, the outside of which hardens to protect the developing embryos within. These cocoons then hatch after a time that depends on soil temperature and moisture content. In temperate regions most cocoons are produced in either spring or autumn and can then develop over the dry summer or cold winter months, whereas in the tropics there is naturally less seasonality in reproduction. After development the hatchling earthworms, sometimes called wormlings, break out of the cocoon from the points where the cylinder originally closed. The number of emerging hatchlings is a function of the species concerned and may range from one to as many as six. The size of the cocoon is also a function of the size of the clitellum of the adult that produced it, so this too is species specific.

The description above may be considered as the standard type of hermaphroditic earthworm reproduction, but as with so many situations in the natural world, other possibilities have evolved. One of these is parthenogenesis (embryo development without fertilization). Some relatively common earthworms, such as *O. cyaneum*, reproduce in this way, effectively producing clones of themselves – a form of asexual reproduction. This must be advantageous under certain circumstances, hence its evolution. Cocoon production is as described above, except that no fertilization with collected sperm is required and no energy is expended in either finding a mate or copulation.

Due to their soft bodies, there are very few earthworm fossils. In 2020 a new fossil crown polychaete (bristle worm) was found from the early Cambrian period (500 million years ago).[28] Also in 2020 simpler fossil worms that pre-dated these by 50 million years were extracted from rocks in South Australia.[29] Nevertheless, earthworm fossils remain tantalizingly elusive. One very rare find from a pre-Bronze Age level (dated at 15,000–8000 BP) represented a fossil earthworm embryo in a cocoon.[30] Even though fossils of the animals themselves may not be found, ichnofossils (trace fossils) representing attributable biological activity may persist. Discovery of *in situ* terrestrial fossil earthworm burrows from just above the impact-defined Cretaceous–Palaeogene boundary (66 million BP) have occurred in North Dakota, USA.[31] The criss-crossing networks of horizontal burrows occur at the interface of lignitic coal and silty sandstone and reveal intense faunal activity within centimetres of the boundary clay. Estimated rates of sedimentation and coal formation suggest that the burrows were made less than 10,000 years after the end of the Cretaceous impact. The burrow characteristics are consistent with burrows of extant earthworms. Another earthworm trace fossil is that of an aestivation chamber discovered from the Upper Pleistocene in Uruguay.[32] From paleosols of the Sopas Formation, these spherical subterranean chambers represent resting places for some earthworm species during adverse soil conditions and exactly mirror those produced by extant species.

The lack of fossils mean that annelids are currently underused in palaeoecological research. However, their DNA might be preserved in sediments and has been reported as an extracellular bycatch from the late glacial and Holocene sedimentary DNA records gathered from the Polar Urals and northern Norway.[33] These suggest that both the soils and the lake sediments remained biologically active over the last 24,000 years, including during the Last Glacial Maximum or the Late Glacial stadials, when cold, dry conditions prevailed and the vegetation was predominantly tundra steppe. Perhaps through environmental DNA (eDNA) extraction, there is still great scope to learn of earthworm activity from the past.

The ancient Greek author Aristotle reputedly described earthworms as 'the intestines of the soil'. This would seem to suggest that Aristotle considered their very presence to be a vital component in soil nutrition and ecosystem function. However, closer inspection of translations now calls this proposition into question, and it is uncertain exactly what type of worm(s) Aristotle was referring to. One translation is given as: 'For all of these [animals], though they have but little blood by nature, are nevertheless sanguine, and have a heart with blood in it as the origin of the parts; and the so-called entrails of earth, in which comes into being the body of the eel, have the nature of a scolex.'[34] This broad description clearly refers to a mix of various types of worms and vermiform creatures and suggests that the ancient Greeks may not have been fully informed about earthworms, and indeed may have thought that there were actual relationships between the different animals, parasitic flatworms and even eels! Hopefully such thinking can now be dispelled. In the next chapters, details

Fossil marine 'worms': an artist's impression of how they may have lived in the Silurian period about 435 million years ago. Described in 2019 by T.A.M. Ewin et al. in the *Journal of Paleontology*, these worms are really a type of echinoderm.

A European blackbird (*Turdus merula*) with a beak full of 'worms' ready to feed to its nestlings.

of earthworm ecology and their activities in the soil are explored more fully and one well-known species, *Lumbricus terrestris*, is given even closer scrutiny, but before that, our own relationships with earthworms will be examined.

2 People's Interactions and Perceptions

We are all likely to have encountered earthworms, whether in a passive setting or, for some of us, in a very active way. Few will have gone through life without ever having had an interaction with earthworms at some point, as they are present wherever soils are found. An experience shared by many is seeing earthworms on hard surfaces after a period of rain, which has led to some species or the whole group being referred to as 'rainworms' in many countries. An old superstition was that these worms would 'rain down', but it is more likely that they 'rain up', emerging from within the soil after rain. Often thought to be a reaction to avoid drowning in their burrows after heavy rain, it's more conceivable that earthworms are taking the opportunity to disperse over favourable, moist soil conditions, as earthworms can survive relatively contently in water and even obtain oxygen from the water through their skin. Those found after rain on concrete surfaces are the ones unable to burrow down. Many earthworm species can be encountered in this way and, should one be interested, this affords an opportunity for collection and observation.

Very young children are often fascinated by living organisms, but many of the creatures they are drawn to can move too rapidly, are too delicate to be handled or may even pose a threat through potential bite or sting. This is not the case with earthworms, which makes them a potential source of great interest; they appear 'alien' by comparison with many other animals, such as pets encountered in early years, yet are safe and no more harmful than any other relatively large creature found in soil. Microorganisms can be a problem to health, so always be prepared to wash your hands after soil investigations. Some children can be particularly fond of earthworms. At primary school I had a friend whose young sister was known affectionately as 'worm-eater' – you may be able to guess that some of her games in the garden led to the consumption of more than mud pies.

Throughout science education, earthworms have featured as prime examples of annelids within the animal kingdom. In higher education, earthworms have repeatedly been a focus for studies of anatomy and locomotion, as whole animals, or for examination of gut components, excretory organs and the circulatory system viewed via sectioned slides under a microscope. Perhaps more interestingly, earthworms have been studied in primary and

A fascination with earthworms.

An angler's tin box containing earthworms collected as bait for fishing.

secondary schools for decades. For example, a discussion of 'The Earthworm' in a section on 'Animal life in the garden' in a science textbook written before I was born contains some very good information. If attentive in their schooling, pupils from the 1950s could have learned much of earthworm ecology and behaviour that would not be disputed today. The idea of creating a 'wormery' by filling a thin, glass-sided container with alternating layers of sand and soil would permit simple but practical observation of activities that bring about soil mixing.[1] It is unfortunate, however, that most of these schoolchildren will have come away from their studies with a view that there is only 'the earthworm', a single species, if indeed the concept of species was considered, rather than the thirty or so that are known to be present in Britain – not to mention the more than 6,000 worldwide.[2]

Many more of us will have become acquainted with earthworms from the experience of simply digging in soil. Gardeners frequently uncover earthworms when weeding or preparing areas for planting and the numbers found may be a function of how the garden is tended, since additions of organic matter, which forms the basis of their diet, attract the earthworms and therefore encourage population growth. Some birds, such as the European robin (*Erithacus rubecula*), have learned of this and become quite tame around active gardeners so as to reap the reward of an easily sourced meal. Similarly, chickens are notorious for catching worms as the soil is dug and will even risk their necks as the spade falls to be first to grab and instantly devour a juicy earthworm: if you are an angler and earthworms are your bait of choice, avoid digging in soil where chickens may be present, as you may toil long for little reward. Fishing is a pastime that may also lead to early encounters with earthworms, either collected from soil, purchased from stores or, these days, bought via the Internet. Anglers probably view earthworms alongside maggots (fly larvae) as eminently disposable and only a means to catch fish. This, in my view, is a sad end for many earthworms that can and do offer so much more. For example, one way that we make good use of earthworms is in the breakdown of organic (carbon-rich and once living) material such as dung or plant food waste.

If heaped in large quantities, most organic materials will naturally start to compost because of the activities of a suite of microorganisms that feed on these substances and in doing so produce heat. A good example of this might be a steaming heap of dung in a farmyard, possibly seen during a walk in the countryside on a cold winter's day. This is one way that organic waste, which is also a valuable resource, can be processed for reuse as a fertilizer, using naturally occurring microbes. This thermophilic (heat-loving) process may reach temperatures of 70°C (158°F), which, although tolerated by specific microbes, would kill all earthworms. However, there is an alternative technology that can be employed to process organic wastes, something referred to as 'vermicomposting', which as the name suggests is a form of composting using worms. This takes place at ambient temperatures and is driven by the earthworms concerned, rather than microbes alone.

A small pile of organic waste, probably less than a cubic metre, will attract 'compost worms': no one is exactly sure from where these originate, but they are often associated with human habitation. Probably drawn from the surrounding area by olfactory cues, small, highly mobile and usually brightly coloured species colonize and begin to consume the organic material. In Europe and the USA these are likely to be the 'tiger worms', 'brandlings', 'dendras' or 'redworms' often referred to in associated literature. Nevertheless, other species that perform the same role may be found in more tropical climates with higher temperatures, but only where moisture content is adequate. Vermicomposting is an excellent way of reducing volumes of waste organic materials but will only work when certain environmental parameters are within given bounds. These include water content, temperature of 16–27°C (60–80°F), pH (around neutral), low salt content and sufficient oxygen. Over a period of weeks these worms aerobically process the waste to produce 'worm-worked material' – their castings, also referred to as vermicompost, which proves to be an excellent horticultural growing medium and has also been shown to contain

A tiger worm used in processing (vermicomposting) of waste organic material.

plant growth hormones, which further increases its desirability as a growth substance.[3] At the same time, the numbers of worms increase until the organic matter is exhausted, at which point the worms will disperse.

An Internet search for vermicomposting or vermiculture, the deliberate population growth of worms, will produce dozens of links to businesses and to publications. The former may stem back to the 1970s when some enterprising individuals began to realize the financial potential of earthworms and started what has developed into a large business venture. The early earthworm entrepreneurs in the twentieth century, many based in the USA, saw the potential for producing worms as fishing bait. The earthworm species used are r-selected, which means that they breed very rapidly, and can be harvested easily from piles of organic matter or from specially designed 'earthworm beds' before being sold to anglers. However, many of these small worms, such as the tiger worm (*Eisenia fetida*), are renowned for producing a noxious, fetid-smelling coelomic fluid when irritated, for example when being put on a hook. These worms were certainly not the best bait for fishing and, although still sold for this purpose, a market for larger soil-dwelling earthworms was more desirable. However, earthworm farmers soon discovered that the worm-worked material might be more valuable than the worms themselves and perhaps that two products, worms and vermicompost, could be obtained from organic wastes by harnessing earthworm activity.

Much has been written about vermicomposting, but some of this writing is not so illuminating and often ill-informed. However, a few quality books for general consumption have been produced. One of the first from the 1970s, *Earthworms for Ecology and Profit*, paved the way for vermicomposting to be thought of as more than a 'hippy pastime'.[4] It was soon followed by Mary

Appelhof's *Worms Eat My Garbage*.[5] Mary had a real passion for vermicomposting and was an enthusiastic communicator of the science. She entranced the general public and scientists alike during the 1980s and '90s and her book is still a best-seller.

For vermicomposting, earthworms are kept by people not as pets but to be farmed as active workers, even if at a household scale. They may be considered as ecological tools able to process organic household waste through vermi-processing. By feeding peelings and other vegetable waste to earthworms, the materials do not become part of the domestic waste stream that winds up in landfill, where it then produces landfill gas and leachate. This therefore removes or reduces a requirement for additional infrastructure, such as leachate treatment and landfill gas collection, to deal with these by-products. Instead, the organic waste can be processed by earthworms at a household level and produce the worm-worked material for growing plants. This may then obviate the sale of peat-based potting compost for household and garden use and assist with the circular economy – a system aimed at eliminating waste and the continual use of finite resources, such as peat.

Many local authorities in the UK now operate waste collection schemes that oblige householders to separate the wastes at source. In this way, many recyclables (metals, glass, paper and card) are collected weekly or bi-weekly along with 'green waste'. The latter usually comprises grass cuttings, hedge trimmings and other plant-based materials. Food waste is not normally deemed acceptable and is required to be disposed of into the standard waste stream. This may seem inappropriate, but locally, authorities often seek to aerobically compost the green waste and wish to avoid materials that may attract vermin such as rats. One way to overcome this dilemma would be to offer householders the opportunity to vermi-compost their own organic wastes, both green and kitchen. As with so many ideas, this is not novel. Adur District Council, in

West Sussex, pioneered such a domestic scheme in the 1980s by providing residents with tiger worms. Accounts suggest that a few residents continue to operate this small-scale vermicomposting operation, but the pilot scheme was never developed fully. Nevertheless, other authorities across the UK also promoted and investigated the use of home vermicomposting. For some, this was through the provision of subsidized vermicomposting bins to residents or simply by promotion of home use, along with details of where bins could be obtained.[6] Although admirable by way of suggestion, detail relating to the extent of how much uptake occurred and the longevity of use could not be easily determined and was unfortunately thought to be minimal. There is still so much that can and perhaps ought to be done at a household level to have worms work to process waste organic material.

In addition to processing vegetable waste products, vermicomposting can also tackle animal dung and human waste. Research in the USA during the 1980s looked specifically at harnessing compost worms to deal with sewage sludge (biosolids). Considerable efforts went into finding the best species to use and the ideal environmental parameters for the process. All appeared to work extremely well, but scaling-up to an industrial level was more difficult and has still not been fully developed. The difficulty is that this process has living organisms (the worms) at its centre and, unlike microbes, which are the biological agents in standard sewage treatment works, earthworms are more susceptible to environmental influences – whole populations could die within a treatment system if the temperature, pH or oxygen levels shift drastically or even through the accidental introduction of a toxin. Nevertheless, worms can be useful on a small scale in processing

Opposite page, clockwise from top left:

A 'Can-o-Worms' household wormery, with sections through which the worms process organic materials.

Upper wormery layer where organic waste such as kitchen peelings are deposited.

Vermicomposting of organic waste material.

Worm-worked material (vermicompost), a product of epigeic worm activity and a useful plant growth medium.

Vermiprocessing of household organic waste in action.

human waste, as is seen in the use of compost toilets in some remote areas, where mains sewage connections are not available.[7] Here, worms are easily able to cope with the quantities produced. In some developing countries, it has been suggested that the use of compost toilets, sometimes referred to as 'tiger toilets' because of the worms involved, could greatly improve sanitary provision for a huge proportion of the world's population. Although difficult to quantify, it is believed that much of the world's organic waste goes untreated, with an estimated 350 million tonnes of untreated human faeces annually discharged into the environment. Research has shown that in countries such as India and South Africa, availability of composting worms is not the main barrier to this and that the viability of introducing such technology exists. Worm provision could be met by existing suppliers with the potential for export to neighbouring countries too. There is an urgent need for the adoption of sustainable organic waste-processing technologies and the use of proven vermicomposting at appropriate scales could certainly meet a large part of this.[8] Such systems would also reduce potential helminth-related diseases.

Although most of us might regard earthworms as beneficial organisms, sports turf managers regard earthworms as hindrances to their work. This is because the natural activities of some soil-dwelling species produce worm-casts, which end up on the soil surface. These castings mar sports fields, greens and lawns, where the smooth running of a ball on a grassy surface is required. For example, a championship bowling green, a grass-surfaced tennis court and the greens on a golf course are not places where earthworms are appreciated. Turf managers seek to eradicate earthworms to prevent the production of surface casts and use poisons for this purpose. In an amateur gardening publication from 1915 it was suggested that a remedy for worm-casts produced

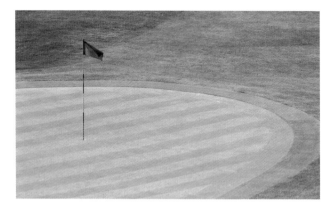

Managers of sports turf such as golf greens abhor earthworms, as the smooth rolling of a ball can be impeded by any casting deposited at the soil surface.

by these 'vexatious' animals was 'a dressing of sharp, gritty sand [which] sends them to the lower strata'.[9]

Mole catchers also once used earthworms as part of their arsenal of weapons to remove the hill-creating mammals from lawns. Strychnine-laced earthworms were put into the mole tunnel to tempt the occupant to dine on a poisonous bait. Strychnine, however, was withdrawn as a method of mole control in 2006 and now artificial 'worms' with an inbuilt toxin can be used to control moles that may appear in private gardens or on large, high-value amenities such as golf courses.[10] Although worms have been poisoned and used as a vehicle to deliver poisons to other organisms in the past, this now thankfully occurs less often for the sake of the wider ecosystems and their living components, but the earthworms are the real winners. Alternative earthworm exclusion measures (behavioural, acoustic and vibrational) for sporting turf are now in demand, creating yet another potential area for future research.

In contrast to a view of earthworms as 'enemies' in turf grass, to some people earthworms can act as the focal point of entertainment and competition: annual worm charming competitions are held on the fields of a local school in Willaston, a village in

European mole (*Talpa europaea*) devouring a lobworm. These mammals collect and store immobilized earthworms in an underground larder.

Cheshire, and near the Normandy Arms Public House at Black-awton, a hamlet in South Devon, in the spring or early summer of each year.[11] Both events date back to the 1980s and encourage the general public to extract as many worms as possible from an area of grassland, also known as a worm charming plot, in a given time, although rules vary slightly over the time allowed and size of the plot. The events at Willaston and Blackawton, as well as in other villages, are very popular and each claims to be the original worm charming event. Other than specific local rules, which can vary, the main aim of the competition is to coax earthworms from the soil without digging into the grassy surface. Liquids can be applied to the worm charming plot, but only those that the contestant is prepared to drink. Common extraction methods include sending vibrations into the soil by wiggling an inserted garden fork, jumping up and down, or playing music to vibrate the turf. Whatever rules are applied, the numbers of earthworms extracted are totalled at the end of the competition, a winner is declared and the worms are returned to the soil. No earthworms are harmed during the worm charming and money is normally raised for charities from entrance fees. A research colleague and I thought about entering one of these competitions and using our collective ecological earthworm knowledge to try and bring home the title of 'Worm Charmer of the Year'. Over more than two decades of collaboration, however, we never managed to enter a worm charming event, possibly because we were never in the right place at the right time. Secretly, though, I know that the reasoning behind this is that we were aware we could very easily lose to an eight-year-old and their grandparents who have no earthworm expertise whatsoever. Perhaps it is best for some to steer away from events like this and contemplate the heterogeneity of earthworm distribution in grassy areas and the luck of the draw when it comes to allocation of a worm charming plot!

It should be noted that most of the methods used in worm charming are very sensible and can cause earthworms to surface. The vibrations sent into the soil may mimic the action, sometimes called foot trembling, of many bird species, including gulls and

The annual World Worm Charming Championship at Willaston in the UK. Competitors seek to entice the most earthworms from a grassland plot in a limited time frame.

waders,[12] and even the similar stomping action of some reptiles such as wood turtles.[13] All are proven ways of encouraging earthworms to surface and in these cases provide a meal for the hungry predator. These mimic the action of rain hitting the soil surface, which is known on occasion to bring earthworms from their burrows as they seek to colonize new areas, as previously described.

Historically, unlike birds, mammals and larger insects, which are generally very showy and easily observed, earthworms were not studied as extensively. Victorian gentry would avidly observe and collect specific components of natural history, such as bird eggs and the birds themselves (for taxidermy), or fill drawers with pinned and labelled insects. This was not the case with earthworms, which, save for Charles Darwin, had very few champions. For this reason, unlike the more popularly studied animal groups, distribution maps, even at a county level, are almost non-existent for earthworms. Only organizations such as the Natural History Museum in London and specific soil ecologists held any records of this nature. It is relatively recently, since the start of this millennium, that we have begun to increase our knowledge of earthworm distribution – the basic 'where' in ecology – at a county and national level.[14] The general public has been recruited to assist in this area through citizen science projects, the involvement of the public in research, which can grow scientific capacity and increase public understanding of science. With earthworm distribution, this has been partially achieved by the creation of more user-friendly identification guides such as that from OPAL (OPen Air Laboratory) and Riverford Organic Farm's 'Big Worm Dig'.[15] Although not perfect, with some assistance and follow-up examination of collected earthworms, a better understanding of earthworm distribution in Britain is forming and being assisted and even driven by the work of the Earthworm Society of Britain.[16] This is a gradual process that will build over time and one that is also being undertaken in other countries. By contrast, so-called BioBlitz events can also be valuable: in a short space of time, for example, 24 hours, individuals including academics and amateur naturalists with expertise in biological identification are brought to a nature reserve, park or other location and asked to identify as many species as possible to help build a greater knowledge of the biodiversity of that area. A dozen or so earthworm species are often found within a few hours using tried and trusted extraction methods. This has, at times, equated to a significant contribution to the overall number of plant and animal species found and identified in the given area of the BioBlitz event.

We have already mentioned collection of earthworms after heavy rain or by digging (avoiding chickens, that is), but what if those sought are of a particular type? In this case, perhaps it is better to employ a variety of techniques, although rarely is there a need to use the 'grab and hold' technique as described for *Lumbricus badensis*. Looking beneath stones or other material on the soil surface may be productive, but standard collection usually involves digging within a small area of soil (¹/25 or ¹/10 square metre) to a depth of around 20 centimetres (8 in.) and examining

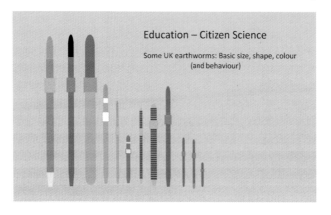

Education – Citizen Science

Some UK earthworms: Basic size, shape, colour (and behaviour)

Common British earthworms drawn to aid identification for a citizen science exercise.

worm identification key

Use this key to identify which ecological group each worm belongs to.

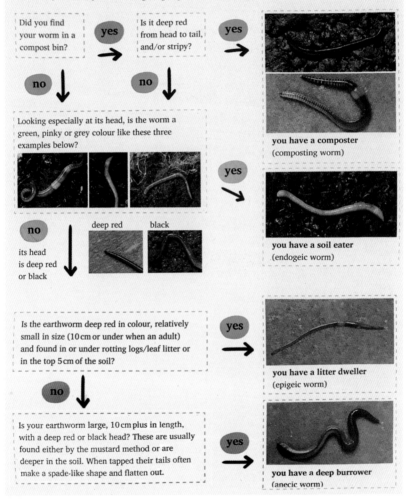

Did you find your worm in a compost bin?

yes — Is it deep red from head to tail, and/or stripy?

yes → you have a composter (composting worm)

no ↓

Looking especially at its head, is the worm a green, pinky or grey colour like these three examples below?

no ↓

yes ↘ you have a soil eater (endogeic worm)

its head is deep red or black

deep red   black

Is the earthworm deep red in colour, relatively small in size (10 cm or under when an adult) and found in or under rotting logs/leaf litter or in the top 5 cm of the soil?

yes → you have a litter dweller (epigeic worm)

no ↓

Is your earthworm large, 10 cm plus in length, with a deep red or black head? These are usually found either by the mustard method or are deeper in the soil. When tapped their tails often make a spade-like shape and flatten out.

yes → you have a deep burrower (anecic worm)

the soil for earthworms by hand (hand-sorting) on a plastic sheet.[17] This will find those (epigeic) earthworms that live in organic matter at the soil surface plus those (endogeic) that have relatively shallow burrows. To extract the deeper burrowing (anecic) earthworms it is necessary to add a vermifuge, a liquid that will enter earthworm burrows when poured into the pit already dug and act to irritate the skin of the worms, driving them upwards. These days a standard method is to either use a suspension of mustard powder in water or allyl isothiocyanate, the active ingredient in mustard. Worms extracted in this way will be unharmed if they are washed as soon as they emerge. This then permits the abundance and biomass (number and weight of earthworms per square metre) to be calculated. After identification using a standard guide, the species richness can also be determined for the given sampling area.[18] In areas where digging may not be permitted, for example in some nature reserves or on golf greens, another technique may be used: earthworm extraction using electrical stimulation. By inserting electrodes into the soil and attaching them to

A simple identification guide used by Riverford Organic Farm's 'Big Worm Dig' to subdivide earthworms into ecological categories.

A combined (standard) method for collection of earthworms. This involves searching by hand through soil dug from a pit of known size and then application of a vermifuge to the same area to allow collection of all earthworm ecological groups.

Delivery of a measured electrical current into the soil can be used to collect earthworms, where damage to the local habitat through digging needs to be avoided.

an appropriate device, such as the Octet apparatus developed in the 1980s in Germany, worms can be coaxed out of the soil by passing a pulsed electrical current between the electrodes.[19] The efficacy of the technique has been investigated and, while providing similar information to both hand-sorting and applying a vermifuge, results may be influenced by soil moisture content. Earthworms can be caught, but can more be done with them than increasing scientific knowledge?

Earthworms have long been used as a source of protein by humans. Their inclusion in the diets of indigenous South Americans was recorded by Alfred Russel Wallace more than a century and a half ago.[20] Some native groups regarded worms as highly desirable and nutritious food, while others did not. Those that ate earthworms distinguished different kinds that were collected in different habitats, such as the wet soils beside streams and from within soils of the highland forests. These were processed before eating by being heated in water up to 80°C (176°F) or smoked over a wood fire. Prior to consumption, the worms would be manually stripped of the gut contents or, if a larger size, cut open lengthwise to remove any soil material. More recent research has demonstrated that the protein content of an earthworm is high, at around 60–70 per cent dry mass, and contains many essential amino acids and minerals.[21] It is perhaps no wonder that many Indigenous people regarded such animals as a delicacy and were prepared to go to great lengths to obtain them.

There are reasons why earthworms may not become a regular dietary option for humans. Accumulation of potential toxins, such as heavy metals, from the organic substrate on which they may be grown could be a problem, although experimental work with fish and chickens seems to suggest this may not be so apparent. Earthworms acting as an intermediate stage in transmission of parasites, such as nematodes and protozoans, is another consideration. Care would be needed here to prevent any potential transmission of viable eggs or cysts of any parasites to humans. If these conditions were avoided, it would seem that earthworm burgers or other earthworm-derived foodstuffs would be nutritious for humans as a luxury food. If the economics could be squared, then earthworms could prove to be of value and assist with feeding a growing global human population. In that case, however, perhaps the biggest hurdle for progress might be human psychology: the revulsion many people instinctively have at the thought. In the Far East, for example, many types of insects are readily consumed and across the world myriad crustaceans, such as shrimps, prawns and crabs, are eaten. Why not worms? A colleague of mine, now retired, once furnished me with a delicious paté made from earthworms. Flavoured with a little garlic, it was as good as any shop-bought equivalent made from more traditional meats. Couldn't meat-eaters consider a move in this direction? The critical thing when using earthworms as a food source is to ensure that their gut is fully emptied, as evidenced by the Indigenous Americans who did this manually. A simpler way is to encourage the worms to empty their own gut contents (depuration), which can be achieved on a small scale by keeping worms in a container with nothing but damp tissue paper for 24 hours. Thereafter, the simplest and kindest way to kill fresh worms is by freezing. Many recipes using earthworms have been published and an Internet search will find a good selection, but earthworm recipes are not new. A national earthworm recipe competition was held in 1975, sponsored by the California-based company North American Bait Farms, seeking ecological gourmets and, no doubt, publicity. Finalists' entries included an omelette, a curry and patties (burgers) but the winner was 'Applesauce Surprise Cake', by entrant Patricia Howell from St Paul, Minnesota.[22] Quite a surprise, I'm sure!

The drive to commercialize vermiculture in the 1970s, as a way of producing earthworms for the fishing bait market and vermicompost as a horticultural medium, took another turn when it was envisaged that worms could also be a useful source of protein. If not directly for humans, then this could act as a substitute for fish meal or meat meal to be fed to poultry, pigs or fish within the intensive livestock industry, or even as pet food. The nutrient content of worms was not seen as a problem, but it was found that the economics did not stack up. Scaling up from an experimental basis seemed to indicate that earthworm protein was as expensive to produce, or more so, than fish meal. (Nevertheless, it is worth noting that with dwindling fish stocks in the twenty-first century, perhaps it is time to reappraise this situation.) The earthworms produced, even though of the composting variety, were more valuable when sold as bait or to start more worm farms, in a form of pyramid selling. Calculations from the 1980s put the cost of fresh earthworms at more than that of prime beef steak. The notion of pyramid selling is not one that will be dwelt upon here, but if some well-meaning person offers you shares in a worm-farming business, in a deal that seems almost too good to be true, then think very long and hard before agreeing to part with your money.[23] Acting as an expert witness in legal disputes, I have seen the unfortunate outcome from several cases.

Earthworms also feature in some forms of healthcare, such as Traditional Chinese Medicine (TCM), in which many claims to the health benefits of earthworms are made.[24] In TCM the earthworm is sometimes called the 'earth dragon' and has been advocated to improve blood circulation, to treat strokes and as antipyretic and diuretic agents. Whether improvements have been brought about by use of earthworm extracts in these instances may be open to interpretation, but it is of some interest that the potential of earthworms is considered vast. Over many years I have witnessed presentations at numerous earthworm-focused conferences on a range of health-related projects. In humans, use of earthworm parts/derivatives are meant to have improved libido, increased blood circulation and even had spermicidal qualities for use as

Earthworms as a source of protein for humans, adapted from M. Paoletti et al., *Proceedings of the Royal Society London* (2003). Nutrient content of earthworms consumed by the Ye'Kuana people of the Alto Orinoco, Amazonas, Venezuela.

A knitted lobworm, created to scale, with a realistic-looking clitellum.

a contraceptive. Claims have also been made that mothers in the postpartum period increased breast milk production after consumption of earthworm.[25] Although some medical research using earthworms may still have considerable unanswered questions, a recent study demonstrated that the coelomic fluid of *Dendrobaena veneta*, a common composting earthworm, exhibited toxicity against human lung cancer cells but not towards cells of the bronchial tubes.[26] This was heralded as a breakthrough by some in the potential treatment of one form of cancer. However, this selective effect on the tumour cells was only achieved after heat pre-treatment of the coelomic fluid at 70°C (158°F) and the effect was time- and concentration-dependent. The researchers suggested that these *D. veneta* coelomic fluid results were an interesting and promising preparation for further biological, chemical and biomedical research, although much hard work may still lie ahead to make this of any real practical significance. Understandably, there will certainly be a need to identify the active components of the coelomic fluid and determine how they operate before testing the clinical potential of this approach.

An unusual animal as perceived by children, an educational resource, an agent for processing waste, a bait for fish, food for humans or animals, and even as a source of materials for health care, earthworms may well impact or have impacted upon the everyday lives of us all. But so far this book has only scratched the surface of the main reason that we are all aware of earthworms. Our overriding interaction with worms is in the soil, where they are often viewed as a gardener's and farmer's ally, and for good reasons. This requires exploration in greater depth.

3 Darwin's Plough

It is believed that worms, through bioturbation activities (natural turnover of substrate), played a vital part in historical sediment mixing across the globe. Through this activity, and the consequent level of nutrients supplied to the soil, they were likely to have assisted in the rapid evolution of marine organisms in the Cambrian period. Descendants of these animals, probably via a semi-aquatic existence in freshwater systems, later brought about similar changes in the terrestrial environment as they evolved into soil-dwelling worms and colonized the land some 200–140 million years ago.[1] This may have been a 'chicken and egg' situation (or more rightly an 'earthworm and soil' situation). What came first? Earthworms are known to be prime agents in soil formation (pedogenesis) so their access to a terrestrial existence may initially have been restricted, until suitable land-based ecosystems had been established. In other words, they needed to facilitate their own environment, and so acted as ecosystem engineers. In the millions of years since, from the Cretaceous period to now, earthworms have become a vital component of soil fauna worldwide and are often the largest animal group in the soil by mass. Indeed, it is often suggested that the mass of cattle grazing in pasture may be less than the mass of earthworms found in the soil below. This chapter will examine the effects that earthworms have in soil, how their perceived role has changed with

'Darwin's plough' exposed to predation through less natural tillage. Image from the works of Lawrence Bradshaw on the bronze doors of Cambridge Guildhall.

time and how one of their greatest supporters was a very well-known scientist. Some of his findings will be examined and comparisons drawn with current research.

Today, most of us would regard earthworms as a benefit to soils, but in Europe until the early nineteenth century earthworms were regarded with some disdain. It was written in a gardening encyclopaedia that, 'Of worms (class Vermes, L.), there are only a few genera which are materially injurious in gardens, the earthworm (*Lumbricus*), the slug (*Limax*), and the snail (*Helix*).'[2] Earthworms were therefore assigned by gardeners to a collective of noxious creatures, as it was thought that they and our well-known gastropods ate the roots of cultivated plants. The gardening advice continued with: 'Saline substances mixed with water are injurious to most insects with tender skins, as the worm and slug; and hot water, where it can be applied without injuring vegetation, is equally, if not more powerfully, injurious.'[3] This advice was followed more directly with:

> The earth-worm is most effectually kept under by watering with lime-water. Salt, vinegar, alum, or other acrid waters will have the same effect, but are injurious to vegetation, and besides less economical. The lime-water . . . is to be prepared by pouring water on quick-lime, and letting it stand till it settles clear, the ground infested with worms should have their casts scraped off, and then the water should be applied from the rose of a watering-pot. The evening, and early in the morning, or on approaching rain, are the best season.[4]

It is apparent from this description of earthworm eradication from areas where they formed an 'infestation' that they were considered as soil pests.

It was not until details of earthworm activities became fully known that a reversal in attitudes towards these animals and their positive effects in the soil became more widely accepted. In his famous *Natural History of Selborne* Gilbert White wrote: 'Earthworms, though in appearance a small and despicable link in the chain of nature . . . if lost, would make a lamentable chasm.'[5] Having noted the existence of feeding relationships and the importance of worms within these, he continued:

> Worms seem to be the great promoters of vegetation, which would proceed but lamely without them, by boring, perforating and loosening the soil, and rendering it pervious to rains and the fibres of plants, by drawing straws and stalks of leaves and twigs into it; and, most of all, by throwing up such infinite numbers and lumps of earth called worm-casts, which being their excrement is a fine manure for grain and grass.[6]

It is quite remarkable that here, from a work published more than 230 years ago, White refers to many activities of earthworms that we, as scientists and gardeners, now take for granted and continue to investigate in greater detail. Although a country clergyman, White was ahead of his time scientifically and was specifically a great observer of many aspects of the natural world within his local surroundings. White also wrote that 'A good monography of worms would afford much entertainment and information at the same time and would open a large and new field in natural history.'[7] He was right, but it would be another century before such a book appeared, and it fell to Charles Darwin to write it.

Charles Darwin's *The Formation of Vegetable Mould Through the Action of Worms with Observations on Their Habits* (1881) totally changed the public perception of this group of animals.[8] This

THE FORMATION

OF

VEGETABLE MOULD,

THROUGH THE

ACTION OF WORMS,

WITH

OBSERVATIONS ON THEIR HABITS.

BY CHARLES DARWIN, LL.D., F.R.S.

WITH ILLUSTRATIONS

LONDON:
JOHN MURRAY, ALBEMARLE STREET.
1881.

The right of Translation is reserved.

Cartoon from the Fancy Portraits series for *Punch* depicting Darwin contemplating earthworm science.

Darwin's 'little book' on earthworms was published the year before his death and contained findings from decades of his observations.

book looked at what we now call the ecology of earthworms, but also explored their potential intelligence.[9] Published in the year before Darwin died, it contained the culmination of a lifetime of dedicated observation on the behaviours of this invertebrate group. Perhaps best known for his work on evolutionary theory,[10] Darwin published a wide range of investigations on other groups, including barnacles, orchids and insectivorous plants. Nevertheless, he held a lifelong interest in earthworms and, as previously noted, doubted that any other animal had played so important a role in world history.[11] This sentiment is not unlike that of Gilbert White from a century before. Famed for his voyage on HMS *Beagle* (1831–6), which included a visit to the Galapagos Islands, Darwin then settled into a home at Down House in Kent in 1842 and lived there for forty years until his death. It was during this period that, among his other interests, Darwin became fascinated with, made observations on and detailed the activities of earthworms.

Darwin recorded the behaviours of earthworms in his monograph, but in contrast to White's 'headline-like' short letter, he detailed and collated observations meticulously, so much so that within his book Darwin describes behaviours and traits that earthworm ecologists recognize even today. Despite not distinguishing between different earthworm species, Darwin was acutely aware that these existed and remarked, 'The British species of *Lumbricus* have never been carefully monographed, but we may judge their number from those inhabiting neighbouring countries. In Scandinavia there are eight species . . . we are here concerned only with the kinds that bring up earth to the surface to form castings.'[12] Darwin was more interested in the collective activities (actions) of earthworms and particularly those that had noticeable effects at the soil surface, rather than seeking to subdivide and pigeonhole the group into different species. From his

The 'Wormstone' at Down House. This was used by Darwin and his son Horace to estimate the action of earthworms on soil turnover.

writings, however, it seems that when he took worms into his house to undertake experiments, he selected but a single species, more of which later.

A colleague and I were permitted by English Heritage to undertake investigations on the estate at Down House from 2004 to 2006. Using Darwin's own writings, we set out to find the exact, or equivalent, locations where his observations were made. From this we then sought to record which earthworm species were present, their abundance and biomass. Overall, in the grounds of the Down House estate and its immediate surroundings, we found nineteen of the earthworm species present in Britain.[13] Some of these were far more numerous than others. Just four species were present within Darwin's Kitchen Garden, for example, but it was here that we recorded the greatest abundance of 715 earthworms, with a biomass of 215 grams (7½ oz) per square metre, values that were relatively high for earthworms in good-quality soils. By contrast, numbers were less than twenty with a biomass of just 12 grams (²/₅ oz) per square metre under coniferous woodland with a layer of undecomposed litter on the surface providing a clear demarcation to a soil that had a poor structure.

Darwin provided some estimates of earthworm numbers per unit area, but most of his figures do not come from systematic sampling. For example, one incident he reports is from disposal of barrels of vinegar (accidently acting as a vermifuge) that led to earthworm expulsion from the soil and their death: 'the heaps of worms which lay dead on the ground were so amazing . . . [I would] not have thought it possible for such numbers to have existed in the space.'[14] Darwin does provide one quantitative figure of '133,000 living worms in a hectare of land' (equivalent to 13.3 per square metre) but this is taken from another published source.[15] We are now much better informed of earthworm abundances and appreciate the effects of soil type, climate,

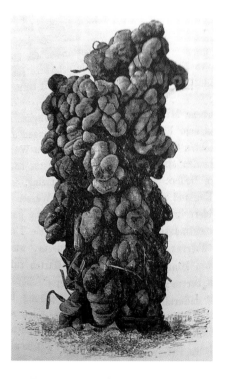

A tower-like casting of a tropical earthworm, adapted from Darwin, *The Formation of Vegetable Mould* (1881).

vegetation and, importantly for the earthworms, a source of nutrition. It has been shown that earthworm abundance can reach over 1,000 per square metre, as found, for example, on part of a reclaimed landfill site. Here at a gate, dairy cattle gathered twice daily prior to milking and regularly deposited large amounts of dung, no doubt acting as a rich source of earthworm food.[16] However, this is exceptional and more likely some hundreds of earthworms per square metre would be the normal abundance for pastureland.[17]

Continuing with investigations based on Darwin's descriptions, we also examined large earthworm burrows that opened at the soil surface. Injection of a mustard vermifuge from a syringe directly into these burrows, which were hidden beneath surface middens, produced only one earthworm species. These burrows were most easily located on gravel-covered areas such as Darwin's Sandwalk (his thinking path) and immediately adjacent to the veranda at the back of Down House, close to an ancient mulberry tree. Although Darwin did not record which earthworm species were present at Down House, from his detailed writings we are certain that many of his direct references are to both *Lumbricus terrestris* (middens, large burrow openings and leaves pulled in) and *Aporrectodea longa* (copious surface casting). The management of the site has changed little over the past century and a half, and the earthworms that we located are almost certainly the descendants of those Darwin observed.[18] Darwin's detailed observations were focused on those earthworms that cast at the soil surface – the deep burrowers (anecics) – and he paid little, if any, attention to shallow-working (endoges) and to litter-dwelling (epiges) earthworms, but collectively these three ecological groups are responsible for a multitude of positive activities in the

Tower-like castings produced by the British earthworm *Aporrectodea longa*.

soil. Of these, some of the most important are turnover of soil, considered as natural ploughing, through the mixing of mineral and organic elements to form nutrient-rich castings, and burrow construction, which permit soil aeration and water infiltration and percolation, so preventing surface water run-off and potential soil erosion.

Earthworms are now considered to be ecosystem engineers – organisms that change the environment in which they live for their own benefit, while at the same time potentially impacting on other organisms.[19] By passing soil through their guts and incorporating organic matter, earthworms have created an environment (through soil formation and enrichment) that suits their everyday requirement. (A more often-quoted example of an ecosystem engineer, and one more easily seen owing to its scale, is the beaver, which can flood areas by damming streams, thus creating preferred wetlands in terrestrial habitats.) The scale at which earthworms operate means that their activities went unnoticed for millennia but were highlighted and initially quantified by Darwin, even though the term 'ecosystem engineering' would not be brought into use for another century. From his observations, Darwin was able to suggest that earthworm casting might be in the range of 2–250 tonnes of soil per hectare per year (0.9–111 tons per acre per year). Recent estimated soil ingestion rates for earthworms in temperate regions are less than 100 tonnes of soil per hectare per year, but easily fall within the bounds given by Darwin, whereas under tropical conditions, with no seasonality, this figure can be ten times higher.[20] Earthworms can process a quarter of the near-surface organo-mineral (Ah) soil horizon on an annual basis and are therefore important aggregate-forming agents through surface and sub-surface cast production.[21]

The very lives of earthworms, no matter what type, are driven by a need to exist in suitable soil conditions, determined primarily

Earthworm casting (faeces) on the soil surface of an experimental grassland plot.

by temperature, moisture and food availability. When soil temperature is acceptable and adequate moisture is present, the only remaining thing that earthworms require is food, which they actively seek in order to grow, survive and reproduce. For endogeic species, such as *Aporrectodea caliginosa* (the grey worm) or *Allolobophora chlorotica* (the green worm), this will require horizontal burrowing through the soil and eating soil along the way. This ecological group are therefore geophagous (consume soil). Many endogeics derive the nutrition they require from the fragments of organic material within the upper layers of the soil and/or through the digestion of the microorganisms that are themselves feeding on this material.[22] Because of this, geophagous earthworms are constantly creating temporary burrows, into which they produce casts as they move forwards. In this way, most soil-dwelling earthworms are 'working the soil' invisibly, that is, without showing any obvious signs of their activity above the surface, with as much as 90 per cent of earthworm casting in pasture underground, so it is effectively ploughed from below.

But how does the organic matter descend from the soil surface where it might accumulate as leaf fall in the first instance?

Anecic and epigeic earthworm species are considered detritivores, feeding on dead and decaying organic material, such as leaves.[23] Epigeics usually live within the leaf litter, but the larger, deep-burrowing (anecic) earthworms feed at the soil surface and actively draw fallen leaves into the burrow they have created, as recorded by Darwin. Measurements have shown that more than 98 per cent of leaf fall, for example in an apple orchard, can easily be incorporated into the body of the soil by the action of earthworms, particularly *L. terrestris*.[24] Leaf litter incorporation of this nature is therefore seen as positive where the earthworms are native to the area, as it increases soil fertility. In addition, deep-burrowing earthworms also cast upon the soil surface, and in doing so can help to bury further surface organic matter. In these ways, deciduous leaves and other organic detritus at the soil surface can become incorporated and more readily available to the geophagous earthworms that may not otherwise have direct access. Species associations of earthworms are regularly encountered in the field, suggesting that niche overlap (a requirement of different species for similar resources, resulting in competition) is avoided and indeed that mutually beneficial behaviours may occur between the ecological categories of earthworms. The life stages of the species may also be of importance here, as juvenile anecic species do not initially produce deep burrows but exist more as endogeics until they reach a given size.[25] At this stage in their life cycle they may compete with adult endogeics and only escape from such a negative interaction once they have grown sufficiently to construct a permanent deep burrow. A host of experimental work has been undertaken in the laboratory looking at inter- and even intra-specific interactions of earthworms, but there is still scope for much research in this sphere of earthworm ecology.[26] When earthworms are investigated in the field, communities of up to a dozen different species can be found living together. This suggests that they may not be competing directly for all resources that they require, and each may benefit in some way from the presence of other species. The subtleties of these interactions, which may well be mediated by environmental factors and by a suite of microorganisms, certainly warrants further investigation and may be of value in assisting crop production in earthworm-rich agro-ecosystems.

In conventionally ploughed agricultural systems, earthworm number and biomass are usually reduced dramatically compared with untilled soils. This is due primarily to the physical damage that is caused to the worms, but also their exposure to avian predators. The sight of birds following a plough has been known for centuries and can still be seen in modern agro-ecosystems. Current agricultural practices using chemical fertilizers and pesticides also have detrimental effects on earthworm populations in most agricultural soils.[27] This is because of the moist skin and soft bodies of earthworms, which make them particularly susceptible to chemical attack. However, within the practice of organic farming, where chemical application is avoided, earthworms are usually present in large numbers and facilitate the more natural processes used for food production and help to improve soil quality through, for example, organic matter incorporation. This is achieved when material such as animal dung or farmyard manure is applied at the soil surface and earthworms act as ploughs through natural bioturbation. A conversion from conventional to organic farming will see a growth in earthworm numbers, but growth of populations may take several years.

To look more closely at the action of earthworms and their positive effects on soils, there is a need to consider the incorporation of organic matter and its intimate mixing with the mineral

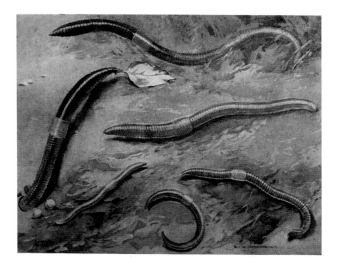

Common British
earthworms,
illustration adapted
from H. Friend,
*British Earthworms
and How to Identify
Them* (1923).

soil in the earthworm gut. This leads to production of castings
that are a rich conglomerate of these components combined with
mucus produced by the earthworm and containing a host of
microorganisms, either from the soil itself or actively from
within the earthworm's digestive system. The 'organic reactor'
that constitutes the earthworm gut differs between ecological
earthworm categories, due to the quantity of soil and/or organic
matter ingested and even differs between species within these
groups. However, the gut effectively has one major function, to
extract the required nutrients from the material ingested and
remove the unwanted material. Food taken in contains a proportion of soil with an aggregated structure. Through physical,
muscular grinding in the gut, the chemical addition of enzymes
and the biological action of microorganisms, the soil is disag-
gregated and thoroughly mixed during the hours of gut passage.
As might be imagined, a host of research has been conducted
on the various elements of this process, from the selection of

This image of a 19th-century oxen-drawn plough shows that birds have always benefited from easy earthworm pickings following soil disturbance.

material by the worms for consumption, the direct action of the gut on these materials and interactions with microorganisms, to the stabilization and nutrient release within castings.

Darwin, through a series of observations and subsequent experiments, examined whether earthworms (*L. terrestris*) showed any preference for the way in which leaves were drawn into burrows.[28] He began by removing leaves from the entrances of inhabited burrows and assessing orientation of insertion, that is, whether the apex was taken first or otherwise. His observations suggested that the majority, and far from due to chance, were taken in by the apex. This was seen, for example, with lime tree leaves and some cultivated plant species. Darwin followed this up by experimentally offering earthworms triangles of writing paper coated with 'raw fat' to prevent them becoming limp when exposed to rain and dew.[29] The pieces of paper were carefully cut into different types of triangles and laid beside earthworm burrows. Inspection the following day allowed for examination of which triangular 'leaves' had been taken and by which point the earthworm had grasped the shape in its mouth. Once again, the results showed that earthworms had tended to take the triangles at their most acute angle, which was easier to access and draw down into the burrow.[30]

The orientation by which earthworms select leaves is important and may suggest a level of intelligence – or does it perhaps show that trial and error are at play? Nevertheless, experiments have shown that earthworms will show preference for different types of leaf litter. This is not new information and research from the 1960s has demonstrated that small circular discs of leaf litter from different trees are taken in a preferential order. Circles are offered to remove any bias caused by leaf size or shape. It is likely that observed preferences are due to the chemical content of the leaves, mainly tannins and phenolic compounds, which reduce palatability and can be detected by the earthworms. By experimentally washing out water-soluble polyphenols, a practice that imitates natural weathering, palatability of oak and beech leaves to *L. terrestris* can be increased.[31] More recent experimental work has involved a cafeteria-style set-up of choice chambers where

Filming of the recreation of one of Darwin's feeding experiments, which investigated earthworm intelligence through provision of triangular paper covered in fat, representing leaves.

earthworms can freely select different types of organic material provided as ground particles of known size in separate tubes below the soil surface. This allows for endogeic in addition to anecic earthworms to be tested, with the remaining mass of contents within the tubes recorded on a regular basis over a period of weeks. This permits choice to be determined, but also which leaf litter the earthworm species switches to after those preferred are exhausted.[32] In a similar manner, earthworms have been tested to determine if they prefer different mixtures of soil and small wheat straw fragments inoculated individually with saprotrophic fungi. All earthworm species tested showed preferences between the six fungal species offered. Early straw decomposers, capable of utilizing water-soluble sugars and cellulose, were preferred in most cases to the lignin-decomposing fungi characteristic of the later stages of decomposition.[33]

Earthworm mucus is composed of glycoproteins, produced externally to keep the worm moist and as a lubricant for movement through the soil or, as relevant here, within the gut as an aid to digestion. The amounts of mucus produced have rarely been estimated, but effects on the activities of microbes within earthworm burrow linings have been studied extensively, producing 'hotspots' of microbial activity.[34] Gut mucus may be added at rates up to 37 per cent in equivalent dry mass to the ingested food in the foregut and much may be reabsorbed in the hindgut, but this mucus acts as a rich resource for the microbes present within the earthworm gut during gut-transit.[35] Digestion is mediated by earthworm-produced and microbe-produced enzymes. Here there is interaction between the earthworm and the microbes to obtain the required levels of nutrition from the ingested materials, whether it is a mixture of soil containing organic matter in geophages or a more concentrated form of dead organic matter with some soil in detritivores. Numerous experiments have been

conducted on the role and fate of the microbes within the earthworm gut, but due to the diversity involved there is no simple answer.[36] Some may be a part of mutualistic interactions, with favourable pH and increased nutrient and water supply swelling their population numbers. Other microbes may be digested, so their numbers reduce and provide the earthworm with valuable nutrition; more may pass through unaltered and may be using the earthworm as a vector to assist in their dispersal. Comparisons of microbial numbers are normally drawn between reference soil taken from the field and from within castings collected nearby. One laboratory-based experiment involved feeding sterilized soil that was subsequently inoculated with plant-health related, forest-derived, soil-borne fungi to three species of earthworm (*Lumbricus terrestris*, *Aporrectodea caliginosa* and *Lumbricus rubellus*) from each ecological category. By collecting casts directly from the earthworms (with a slight squeeze of the posterior segments as required) into small sterile containers, it was possible, through DNA-based investigations and more traditional microbiological agar-plating techniques, to determine which of the inoculated microbes had passed through the gut and which of these were still viable. Results showed large differences between *A. caliginosa* (endogeic species)

Left: the beginning of an experiment to test leaf choice in a 'cafeteria-style' choice chamber, with *Lumbricus terrestris*.

Right: the end of a choice chamber experiment. Tubes around the edge that contained preferred food leaves have been emptied.

and the two other earthworms, but thereafter were very fungus-specific.[37] Earthworm-microbial interactions in soils appear to be complex. The diet of earthworms was a question of some interest even in Darwin's day (see Linley Sambourne's cartoon 'Man Is but a Worm' in the *Punch Almanack* for 1882, which links Darwin's earthworm and evolutionary studies) and this has continued into the twenty-first century. Recent research has now clearly shown that some groups of earthworms, such as the genus *Aporrectodea* and the genus *Lumbricus*, both existing within the deep-burrowing (anecic) ecological category, are very different in the way that they behave, particularly in their mode of feeding.[38] This provides yet another area worthy of further investigation.

Some of Darwin's writings relate to cast production by earthworms, but more specifically to the rate at which these casts are eroded from the soil surface. Not only was Darwin interested in turnover of soil by earthworms, he sought to estimate the rate at which the material brought to the surface might be washed away, particularly on slopes. For a slope of 9° 26', Darwin estimated that about 1,140 kilograms (2,500 lb) per hectare per year of earthworm cast material was removed, which in order of magnitude is similar to mass displacements in major river basins such as the Mississippi, which Darwin was aware of and noted.[39] The contribution of casts to erosion appears to occur following their breakdown by the impact of rain, but there is some debate as to whether more or less erosion would occur in the absence of casts. Some recent results show that surface-deposited casts of anecic species may give resistance to run-off, but others suggest the erosion of cast material leads to a net increase in erosion, with new techniques for such measurement in development.[40] Such findings are from relatively short-term records, whereas over longer timescales of millennia this phenomenon could lead to vast amounts of sediment accumulation in alluvial soil or floodplains.

MAN · IS · BVT · A · WORM.

Linley Sambourne's cartoon 'Man Is but a Worm', which appeared in *Punch*'s *Almanack* soon after publication of Darwin's 'worm book'. The image combines Darwin's theory of evolution by natural selection with his earthworm observations.

One of Darwin's observations on his estate related to a field (Great Pucklands Meadow) that was last ploughed in 1841 and then left unworked to become pasture. He noted that initially it had little vegetation but was thickly covered with small and large flints 'some half as large as a child's head' and was a place his sons called 'the stony field'. He remarked:

> When they ran down the slope the stones clattered together, I remember doubting whether I should live to see these larger flints covered with vegetable mould and turf. But

the smaller stones disappeared before many years had elapsed, as did every one of the larger ones after a time; so that after thirty years [1871] a horse could gallop over the compact turf from one end of the field to the other, and not strike a single stone with his shoes.[41]

Darwin stated that this 'was certainly the work of the worms, for though castings were not frequent for several years . . . some were thrown up month after month, and these gradually increased in numbers as the pasture improved.' In 1871 Darwin measured the thickness of the turf (grass roots) and of the vegetable mould (earthworm casting). The turf was rather less than 1.27 centimetres ($\frac{1}{2}$ in.), and the mould, which contained no stones, was 6.35 centimetres ($2\frac{1}{2}$ in.) in thickness. Beneath was coarse clayey earth full of flints. Darwin determined that the rate of accumulation of the mould during the whole thirty years was only 0.21 centimetres per year (that is, nearly 1 inch in twelve years); but the rate must have been much slower at first, and afterwards considerably quicker. This was long-term ecological monitoring (LTEM) of a slow process that could in no way be hastened, done one hundred years before anyone had even thought of the term LTEM. In a similar vein, Darwin also recorded the burial of layers of cinders and chalk fragments scattered over the surface of the soil in other parts of his estate.

Darwin's records of flint burial in Great Pucklands Meadow led a colleague and I to further our earthworm research at Down House. In 2007 we were given leave to set up an experiment in the same field as Darwin, as we, for the most part, sought to re-create his observations. We gathered flints from a pile collected on the estate and roughly divided them into large and small (approximately 12-centimetre diameter (mean mass 1,100 grams) and 5-centimetre diameter (mean mass 250 grams), respectively).

We then configured a 4 × 4 'Latin Square' field design with these stones presented at either high density (50 or 100) or low density (25 and 50) per square metre, depending on size (four replicates per treatment). In essence, flints were laid out on the soil surface in known locations in the centre of the field where they were unlikely to be disturbed for some years. Photographs were taken of each experimental square metre, so that individual flints could be recognized and followed over time. English Heritage, which managed the site, agreed to avoid driving any mowers or other vehicles over the stones. We have returned to the site three times so far (in 2013, 2017 and 2021), and on each occasion excavated a quarter of each square metre. Our excavations have shown that burial of the flints, as observed by Darwin, has taken place. Our (so far short-term) records suggest that the rate of burial may be approximately 0.73–0.96 centimetres per year (measured over six and ten years respectively).[42] We aim to continue this research ourselves or pass it on to other researchers on retirement, so that a comparable set of data can be collected to that presented by Darwin. It's likely that our own records will differ, certainly over initial years as our flints were placed onto existing turf, whereas Darwin's were on ploughed soil. Equally, earthworm populations will have initially differed due to the previous effects of ploughing. In addition, excavation in Great Pucklands Meadow in 1942 (a hundred years after ploughing) showed that 'Darwin's flints' appeared to have come to rest on flinty clay some 6–8 centimetres (2^1/$_2$–3 in.) below overlying stone-free earth, something that might call into question the rates put forward by Darwin.[43] Our most recent sampling in 2021, which was 150 years since the measurements taken by Darwin and 140 years since publication of his earthworm book, will lead to a further publication.

Earthworms normally move through the soil by exerting pressure using their hydrostatic skeleton. This pushes soil particles to the side and enables the earthworm to move forwards. In addition, under compacted soil situations, for example, all earthworms may eat their way forwards by ingesting soil particles. Either way, they create burrows that, depending on species/ecological grouping, may be horizontal and near the surface or vertical and extend to a depth of several metres. The very act of burrowing creates voids in the soil, with up to 900 metres (3,000 ft) of galleries per square metre that can transmit air, permitting the earthworms themselves to gain access to required oxygen, but also making it available for many other organisms, including plant roots.[44] Some of the deep-burrowing earthworms, such as *L. terrestris*, produce a permanent burrow that they may inhabit for most of their lives. The effect of this structure is to provide a preferential flow pathway from the soil surface into the depths of the soil and, given the diameter of the macropore (burrow), permit rapid influx of water under some conditions of rain and thereby lead to reduced over-surface movement of the water and potential soil erosion. It might be thought that the presence of the earthworm within its vertical burrow would act as a 'stopper' to such water movements, but research has shown that experimental infiltration of water directly into *L. terrestris* burrows could be achieved at rates of 200–300 millilitres (7–10½ fl. oz) per minute.[45] The presence of the worm did not hinder water passage into the soil and may even have assisted it by keeping the burrow open at the surface. The burrows of endogeic earthworms, which are closer to the surface and transient in nature, may permit a slower percolation of water into the soil, which prevents major soil erosion from within and can delay full saturation.

Water is vital to earthworms as they are essentially semi-aquatic organisms. In fact, I once saw a poster at a conference entitled 'The earthworm is a fish!', which went on to explore the similarities between the two groups. Considerable moisture is lost during casting and from cutaneous mucus production, so earthworms constantly extract water from their soil environment and must maintain a moist cuticle for effective gaseous exchange. Earthworm activity and even distribution can therefore be limited by the level of moisture in the soil. Many temperate species are relatively inactive during the summer as soils may be too dry, meaning that during such periods different survival strategies are employed. Some (anecic) species retreat to the depths of their burrows where moisture is adequate, whereas others enter a dry hibernation (aestivation) resting stage. This requires the earthworm to void its gut contents, create a spherical, mucus-lined chamber and coil itself in a knot within. The soil may dry around this structure, but the earthworm is able to survive, as previously mentioned with respect to trace fossils. Other earthworm species are even more dependent on water, and some, such as the European *Eiseniella tetraedra*, are truly semi-aquatic and found to live within the small-sized gravel at the edge of freshwater streams.[46] Adapted to high salt conditions, the euryhaline earthworm *Pontodrilus litoralis* is even known to live in the intertidal region of coastal Western Australia where it feeds on and lives within washed-up seaweed.[47]

As a species, we owe much to earthworms. They deliver many of the ecosystem services that have been hinted at in this chapter, simply by burrowing through soil and eating it combined with organic matter. Major supporting services are offered through nutrient cycling as earthworms recycle dead organic matter and release the materials into the environment to be reused by organisms such as plants, microbes and other animals.[48]

As soil ecologists, we owe much to the work that Darwin undertook during his lifetime. His writing on worms and 'vegetable mould' became a best-seller, perhaps in part due to its perceived quirky content, but also because of the importance it placed on

Top: Outset of part of an experiment to determine the rate of flint burial by earthworms in Great Pucklands Meadow at Down House, 2007, to replicate work of Darwin. *Centre:* Six years after deposition (2013), the flints had started to become buried in the soil, as previously recorded by Darwin at the same site more than a century before. *Bottom:* Close inspection beside a single flint to show the casting action of earthworms.

Aestivating grey worms (*Aporrectodea caliginosa*). The worms coil themselves into spherical chambers, the shape of which helps them to survive under adverse dry conditions.

earthworms and the enormous role that they have had and continue to have in soils. Towards the end of Darwin's book, he stated:

> When we behold a wide, turf-covered expanse, we should remember that its smoothness, on which so much of its beauty depends, is mainly due to all the inequalities having been slowly levelled by worms. It is a marvellous reflection that the whole of the superficial mould over any such expanse has passed, and will again pass, every few years through the bodies of worms. The plough is one of man's inventions; but long before he existed the land was in fact regularly ploughed by earth-worms.[49]

Darwin knew of and made reference to the work of Gilbert White in his book when he stated, 'under ordinary circumstances healthy worms never, or very rarely, completely leave their burrows at night . . . as White of Selborne long ago knew,'[50] but he did not give a full account of the many particularly pertinent earthworm observations of White. This suggests that Darwin may not have wanted to acknowledge that someone had beaten him (by a century) to some very important earthworm-related scientific findings.[51] Darwin and White began our scientific interest in earthworms and because of that and subsequent research we are now better informed on numerous aspects of their lives. However, in the next chapter we will explore how earthworms, and worms in general, are considered away from the sphere of science and have grown more recently to become a part of our culture.

4 Aside from Science

There are many myths, misconceptions and questions associated with worms, not least relating to the existence of the Mongolian death worm or, at a much simpler and perhaps more easily tested level, whether earthworms eat human bodies after death. Or, perhaps the most famous earthworm-related question of all, what happens when an earthworm is cut in half? These and other questions will be addressed here, alongside reasons why worms appear in songs, poems, books and films. As a species we seem to have an unusual fascination with worms. This fascination may be because of the phallic appearance of worms, or it might be a morbid curiosity derived from worms' perceived association with death and decay; perhaps, instead, it is because we simply find worms funny. This chapter will explore how worms are now used as models in engineering and even in computing, where the term 'worm' doesn't necessarily refer to a wriggly, legless animal.

Most people in the West will have encountered William Shakespeare's poetry and plays at some point, whether learning about them at school or seeing the latter performed on stage. Worms were familiar to the playwright and feature in several of his offerings. One of the most famous may be found in *Romeo and Juliet*, when Mercutio, having received a fatal sword thrust, states to his kinsman Romeo, 'They have made worms' meat of me,' indicating that he will die, and by punning with 'ask for me tomorrow and you shall find me a grave man' (III.1). Equally, in *Hamlet*, the prince, when taunting Claudius over where he has hidden Polonius's body, says 'A man may fish with the worm that have eat of a king, and eat of the fish that hath fed of that worm.' This also suggests that worms are associated with death, decay and consumption of the human body and that this flesh will be recycled, 'to show you how a king may go a progress through the guts of a beggar' (IV.2).

I hold the belief that Shakespeare, as with many people who lived four centuries ago, was uncertain of the real action of earthworms in soil and may have thought that, just like maggots, earthworms truly did feast on the flesh of dead people. This is a recurring theme in many art forms, as rendered in the folk song 'On Ilkla Moor Baht 'at', where it is suggested that 'Then t'worms'll come an' eyt thee oop [Then the worms will come and eat you up]'. The song is a warning from a woman, Mary Jane, to her sweetheart that going outside onto the moorland in the inclement

A 'bookworm' astride the complete works of Shakespeare, seen in a bookshop window in Leyland, Lancashire.

Yorkshire winter without a hat (baht 'at) would lead to his death and burial, whereupon he'll be, consumed by worms, the worms themselves will be consumed by ducks, and then the ducks will be eaten by the friends of the dead person, leading to, 'Then us'll all ha' etten thee [Then we will all have eaten you]'. The final suggestion in the song is that, via the agency of worms and ducks, the friends have eaten the dead person. A similar fate is suggested for Hamlet's uncle, Claudius, but Shakespeare's food web consists of worms and then fish, rather than ducks. This may all seem a little far-fetched, but it should be noted that, when referring to earthworms, Darwin wrote that 'Raw fat seems to be preferred even to raw meat or to any other substance which we give them and much was consumed.'[1] This suggests that worms may be partial to animal tissue in addition to leaves and perhaps there is something in the belief that earthworms may 'eyt thee oop'!

As a species we appear to have a fascination with worms and death. A commonly held view, even today, is that when buried, worms will enter a coffin and devour the corpse lying within. For several reasons, a colleague and I were given the opportunity to undertake some investigations in a local cemetery, which enabled us to offer an insight into this.[2] It should be noted that these investigations were mainly aimed at discovering which earthworm species were present in an urban location, and how deep they burrowed, the latter more easily addressed with help from grave-digging equipment. We did locate some deep-burrowing species but found no evidence in the given cemetery that burrows reached down to 1.83 metres (6 ft), the depth at which burials took place. In addition, we considered that the use of a preservative such as formaldehyde to embalm a body would, if earthworms ever reached a corpse, also deter them from eating what they would find to be toxic flesh. It is therefore probably safe to say that, under what for some might be considered standard burial circumstances, our bodies in death do not normally become the food of earthworms.

We have already explored some of the actual parasitic worms that can plague the human body in life, but more than this, worms have historically been perceived as a reservoir of disease and 'having the worm' was formerly a popular name for a collection of various ailments. A good example was the belief that toothache, now known to be caused by tooth decay (caries), was caused by a worm at the root of the tooth and commonly called 'the worm'. Such notions were often fuelled by unscrupulous practitioners who might offer to remove such worms for a fee, and by slight-of-hand produce small worms from investigations into the mouth of the patient and even render some short-term relief by administering a natural painkiller. The hapless patient might then part with money but be little better off in the longer term as 'the worm' was still present. The perceived actions of tooth worms and the types of pain they might inflict are illustrated within an eighteenth-century sectioned molar, 10.5 centimetres (4 in.) high, carved from ivory and known as 'The tooth worm as Hells demon'. The interior of the carving has images of terror and pain inflicted on human-like figures by a rampant worm.[3] Recent micro-imagery of sectioned human molars has shown that worm-like microtubules, much smaller than a human hair, are actually present, but these are concerned with transmission of hot/cold sensations and have no bearing on the perceived tooth worms of old.[4] In a slightly different vein, but still tooth- and worm-related, information extracted from a report that appeared in several American newspapers in 1911 suggested that toothache could be alleviated by adhering to the following advice: 'A worm fed on a particular herb or a cabbage caterpillar can conveniently be placed in a

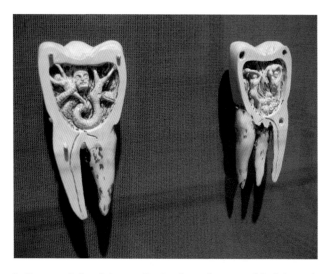

The 'tooth worm as Hells demon': an 18th-century ivory carving depicting the pain of toothache.

hollow tooth, but it is equally simple to chew an adder's heart.'[5] Perhaps this remedy may have given some short-term distraction to toothache, but may not have been very easy to administer.

Within its collection, the Science Museum in London has an Italian tin-glazed earthenware syrup jar that has polychrome decoration and is labelled 'OI LUBRUI', which can be dated to the seventeenth or eighteenth century.[6] The Latin inscription translates as 'Oil of Earthworms'. According to pharmacists of the era this was a pain reliever, especially for aching joints, and recommended for arthritis, rickets and cramp, with the oil probably rubbed on to the patient's skin. Preparation for this treatment was to take 0.2 kilograms (1/2 lb) of earthworms, 0.9 kilograms (2 lb) of olive oil and 57 grams (2 oz) of wine, boil together until the wine evaporated and store for later use. Unfortunately, the recipe does not state if the worms were used alive or dead. It is not known how effective the preparation was, but one writer warned 'they who trust much to it . . . will be disappointed.'[7] Use of such preparations was probably quite widespread, as jars for oil of earthworms are known from other museum collections. This was, at best, ill-informed medicine or perhaps, at worst, another form of quackery, not dissimilar to sales of a panacea called snake-oil in the Wild West of the USA during the nineteenth century. Today, the term 'snake-oil' often refers to deceptive marketing, health care fraud or a scam, although a real product, derived as imagined from snakes, is still marketed and used as a part of TCM.

Now held in the Science Museum (London), this 17th-century 'Syrup jar for oil of earthworms' would have contained a concoction that included earthworms and was used for pain relief.

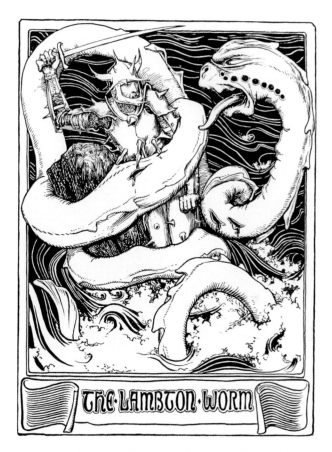

Illustration by John D. Batten depicting the slaying of the Lambton Worm. This mythical creature was said to have grown from a small, worm-like creature into an enormous snake or dragon that terrorized the northeast of England.

Many myths surround worms. Perhaps one of the most famous from Britain is the story of the Lambton Worm, a legendary beast from the Sunderland area in northeast England. The creature was supposedly a worm that grew into a dragon. Caught by a boy called John Lambton who is out angling, the worm at first appears to be like a lamprey (a small eel-like fish), so the boy discards it, throws it into a well and forgets about the creature, but it continues to grow tremendously over a period of decades, poisoning the water source of the local area. The Lambton Worm then emerges from the well and terrorizes the villagers, specifically the Lambton family, by eating sheep, cattle and even babies. As with any myth, there are many versions of events, but the rampage of the worm continued for generations, only ending, it is said, when a descendant of John Lambton finally decapitated the dragon. Although a far from factual account, there appears to be an element of truth to the tale, with several generations of the Lambtons dying under unhappy circumstances. The whole story is told in myriad ways through verse and even in a song, written in 1867 by Clarence M. Leumane, that has become part of local Sunderland folklore.[8]

A more global worm myth has developed over recent centuries and grew with the writing of Roy Chapman Andrews, a respected explorer, naturalist and author who worked at, and later became the director of, the American Natural History Museum in New York. On a zoological and fossil-hunting trip to Mongolia in the early 1920s he was asked by local government officials, including the Premier, to gather any evidence of 'Allergorhai-horhai'. This animal, only known from second-hand reports, was also called the Mongolian death worm. It was described as 'shaped like a sausage about two feet long [with] no head nor legs and it is so poisonous that merely to touch it means instant death. It lives in the most desolate parts of the Gobi Desert.'[9] Chapman remarked that 'to the Mongols it seems to be what a dragon is to the Chinese.' By agreeing to try and capture a specimen 'using long steel collecting forceps and wearing dark glasses', Chapman's expedition to Outer Mongolia was permitted to go ahead. As expected, Allergorhai-horhai was not encountered in the Gobi Desert, although many exciting fossils were found on the expedition, and the death worm may therefore be considered as totally mythical. It was thought

An interpretation of Allergorhai-horhai, or the Mongolian death worm, by Belgian painter Pieter Dirkx, 2006.

that the story may have grown from reports of local desert reptiles, passed down over generations and related to specimens of a harmless snake, the Tartar sand boa (*Eryx tataricus*), which can grow to 1.2 metres (4 ft) in length. Today, the only existence of the Mongolian death worm is in the world of science fantasy. For example, it is featured, with all its attributes, on a card in a 'Monsters of the Mind' section of 'Weird n' Wild Creatures Wiki' and in an action/adventure comedy film entitled *Mongolian Death Worm* (2010) directed by Steven R. Monroe.[10] The latter, very loosely based on the myth, apparently shows what happens when experimental drilling by an American oil company in the vastness of an Asian desert awakens a nest of deadly animals that reproduce quickly and consume all they meet.

Surprisingly, worms have been the focus in more films than might be imagined. Perhaps a whole book could be written about this, but here we will examine just a few. An obvious consideration is *Squirm* (1976), written and directed by Jeff Lieberman, which was billed as the 'night of the crawling terror'. The premise, quite correctly, is that earthworms can be driven to the soil surface by

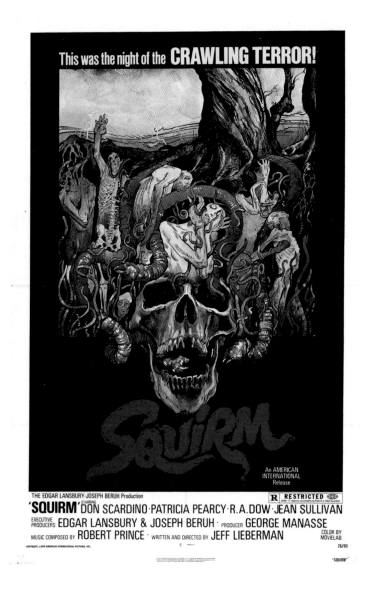

electricity, as happens in the film when storm damage dislodges power cables. The transformation of the earthworms into a tide of flesh-eating creatures, acting under cover of darkness caused by the power failure, leads to a sequence of scenes with increasingly gruesome deaths in the sleepy town of Fly Creek, Georgia. This horror film, using the threat from a well-known but normally innocuous animal, is not dissimilar to Alfred Hitchcock's 1963 film *The Birds*, but lacks the subtle tension and suspense of the classic movie. Rather, it gives the worms an unearthly, unnatural set of behaviours that can be best enjoyed, in pure horror film tradition, if biological understanding of the worms is completely suspended.

Another film in the same genre is *Tremors* (1990, dir. Ron Underwood), which focuses on enormous worm-like creatures that terrorize a town in the heart of the western USA. The subterranean 'monsters' consume livestock and even people, finally finding themselves in a battle with local workmen (played by Kevin Bacon and Fred Ward) who team up with a visiting seismologist (Finn Carter). Some humour exists in this B-rated film, but genuine concern can be felt when the main characters are forced to flee from rock to rock to avoid becoming prey to the Graboid worms that cause small earthquakes as they rapidly burrow through rock-free soil towards potential prey. A final showdown suggests these worms have some intelligence, but they are ultimately outwitted by the humans. The film does not explain the origin of the worms, but some of the characters postulate that they might be mutants (of what?), biological weapons created by the government, an undiscovered carnivorous 'worm' that evolved on earth, or visitors from outer space! My feeling is that this film's portrayal of carnivorous worms has more to offer than *Squirm* and the perceived threat of the 'monster' is made even greater by no direct visual encounter until late in the film's running time. The thought of

The thought of flesh-eating earthworms was probably enough to make every filmgoer *Squirm* during this 1976 horror movie.

falling prey to any living organism, real or fictional, is something that brings real terror to us all.

Dune (1984, dir. David Lynch) was the first attempt to film the novel by Frank Herbert. It features giant sandworms, which may again have been inspired by the Mongolian death worm myth, but here they live on an alien planet. The larvae of these worms produce a substance that is sought after by humans for its medicinal and mystical properties. Obtaining this substance is perilous and forms one of the major themes of a film that is hailed as a masterpiece of science fiction by some and a total flop by others. The first part of another feature film adaptation of the novel, directed by Denis Villeneuve, was released in 2021.

Another obvious film choice would be *The Lair of the White Worm* (1988, dir. Ken Russell), based on Bram Stoker's novel of 1911, which itself was a retelling of the Lambton Worm myth. Russell moulds this further into a more modern setting with sexual intrigue and plenty of horror with more than a hint of vampirism. The film depicts the worm as a giant snake that lives in a tunnel below a mansion and feeds on local animals and people. Except for the alliteration of the title, this might as well have been called *The Lair of the White Snake* as it is not worm-related in any real sense.

By contrast, there appears to be much humour to be derived from worms. Cartoons often feature worms and a Brazilian feature-length animation, *Worms* (2013), has earthworms as the heroes. Here a pre-teen worm is accidentally exposed to the world above the soil and must make his way back home while learning about trust, self-respect and true friendship. Single-frame cartoons of worms in newspapers and other media are also popular and may even reference social media. For example, on meeting his potential partner, a fearsome-looking bird, a worm says, 'You don't look very much like your profile picture.' The renowned cartoonist Gary Larson, who produces the comic strip *The Far Side*,

Poster for *Tremors* (1990; dir. Ron Underwood) that depicts one of the giant Graboid worms that menaces a small Nevadan town in the western United States. The worms, owing to their tremendous size, cause small earthquakes as they tunnel through the soil.

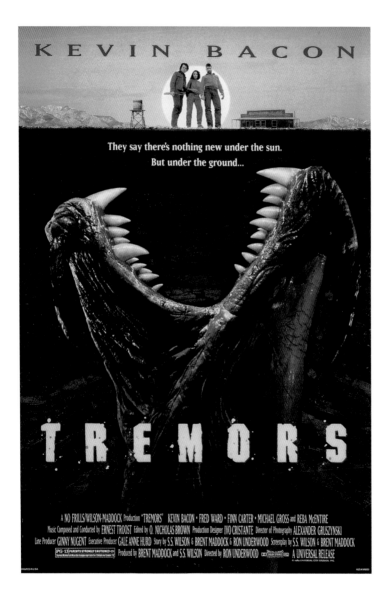

KEVIN BACON

They say there's nothing new under the sun.
But under the ground...

TREMORS

A NO FRILLS/WILSON-MADDOCK Production "TREMORS" KEVIN BACON · FRED WARD · FINN CARTER · MICHAEL GROSS and REBA McENTIRE
Music Composed and Conducted by ERNEST TROOST Edited by O. NICHOLAS BROWN Production Designer IVO CRISTANTE Director of Photography ALEXANDER GRUSZYNSKI
Line Producer GINNY NUGENT Executive Producer GALE ANNE HURD Story by S.S. WILSON & BRENT MADDOCK & RON UNDERWOOD Screenplay by S.S. WILSON & BRENT MADDOCK
PG-13 PARENTS STRONGLY CAUTIONED Produced by BRENT MADDOCK and S.S. WILSON Directed by RON UNDERWOOD A UNIVERSAL RELEASE
© 1989 UNIVERSAL CITY STUDIOS, INC.

certainly shows himself as someone with a firm grasp of the many elements of the biological world. His numerous earthworm-based cartoons draw, for example, upon the outcomes of mistaking one end of a worm for another; a predatory attack from a bird; and the loss of half a worm. These potentially natural occurrences, however, take place during an earthworm party in a house, as if the worms are actors in a horror movie, and as part of a police investigation, respectively. Earthworms are transported into a world where everything that happens to humans can happen to them in a familiar human setting. To my mind, though, Larson's greatest contribution to literature is his book *There's a Hair in My Dirt!*[11] This is, at a basic level, the story of an earthworm family – pipe-smoking Father, lipstick-wearing Mother and their little worm son – who all live together in a tunnel below the soil surface. The premise of the story is about why the youngest family member comes to find his evening meal of a plateful of soil contaminated with a hair. To placate and reassure his son, who rants about being the lowest of the low, Father relates the story of a young woman who lived in the forest and had a love of nature, but perhaps not a full understanding of it. He finally convinces his son that earthworms are really 'Spineless superheroes' and that he should be proud to play his part and continue to eat dirt and fall in with the rest of the living world, which is generally in harmony, perhaps with the exception of us humans. This might be considered quite telling under current global circumstances.

Naturally, earthworms are the focus of many books written for younger children, appealing as they are for that age group. One worthy of mention is *Walter the Worm* (2018), which might not appear to be exceptional, except that this is a new story by Roger Hargreaves (and his son Adam). It is the first of his famous 'Mr Men' books, after 48 predecessors, that does not include the 'Mr' tag in its title. Walter, a yellow-coloured worm, had appeared from his wormhole in many of the previous books as a friend of the 'Mr Men' and 'Little Misses'. This perhaps goes some way to show the attractiveness of worms to children in literature. Without too much of a spoiler, Walter's story relates to an encounter with the 'Early Bird'. Thankfully there is a happy ending.[12] (This might be an appropriate point to mention that although we all know that the phrase 'the early bird catches the worm' means that if you want to succeed, start as soon as possible, in reality it is rather the late worm, straying out on to the soil surface in daylight, that is caught by the bird!)

Roald Dahl's book *The Twits* features, as the title suggests, a couple who are not very intelligent but who like to play tricks on each other. At one stage, Mrs Twit gives her husband a plate of

Selected element of a mosaic from Lytham St Annes, Lancashire, showing a bird about to feed its nestlings with an earthworm.

'wormy spaghetti' to eat. He devours the worms mixed with pasta and covered with tomato sauce, but while complaining of the squishy and bitter nature of his food he still eats the whole plateful of live worms.[13] This may seem far-fetched and totally fictional but has been mirrored in life. In 1991 a trawlerman from Wales ate fifty live worms taken from his compost heap and a handful of lugworms to raise £600 for charity. He was quoted as saying, 'It's just like spaghetti, a quick chomp soon stops them wriggling.'[14] However, the Royal Society for the Protection of Animals (RSPCA) took a dim view of this stunt, no matter how well meaning it was meant to be.

In another well-known book by Dahl, *James and the Giant Peach*, an earthworm has a more central role as one of the animals that James finds inside the very large fruit. This rather quarrelsome

Superhero computer game and cartoon character Earthworm Jim, who is able to perform unheard-of earthworm deeds due to his robotic suit.

earthworm bemoans his routine daily activities, but by the end of Dahl's book it has incredibly foregone 'swallowing soil' to sell face creams in TV commercials.[15] Another cartoon character that will be well known to many is Earthworm Jim, created by Doug Langdale and Doug TenNapel. Originally a character in a side-scrolling computer game that was adapted for a TV series in the 1990s, this earthworm is another superhero. He is able to perform superheroic feats by wearing a robotic suit, so escaping his limb-less earthworm existence. His catchphrase, when overcoming and eliminating those who would do evil, is 'Eat Dirt!'[16]

Surprisingly perhaps, numerous poems feature earthworms or worms in general. For example, 'The Earthworm's Monologue' by Elizabeth Jennings (1926–2001) presents an earthworm speaking about the ups and downs of its own life.[17] This currently features in the online Indian Certificate of Secondary Education (ICSE). Anne Sexton (1928–1974) wrote her sixteen-line 'Earthworm' to include as much about the soil in which it lives as the earthworm itself.[18] Perhaps unsurprisingly, there appears to be little scientific description here but more allegory about the way that we humans treat animals and the environment badly. By contrast, 'The Earthworm' by Donna Word Chappell is a humorous poem that relates how children can interpret things differently to adults: without giving away the punchline, this involves a combination of worms and alcohol.[19] A worm also features in 'A Childish Prank' by Ted Hughes, but in a way that tries to explain how Adam and Eve differ, yet are drawn together for sex, as a worm seems to reunite after severance into two halves.[20] This is quite a different type of poem with no signs of humour, but is perhaps more of an attempt to explain something of mischief-making in the world as brought about and interpreted by a crow. Another corvid bird features alongside a worm in John Gay's (1685–1732) 'The Ravens, the Sexton and the Earth-Worm'.[21] Here the worm revisits and reunites the themes of death, decay and the consumption of dead creatures, whereas 'The Man Who Dreamed of Faeryland' (1893) by W. B. Yeats is a poem that contains, of all things, a singing lugworm.[22] Once again, this ends with thoughts of worms and death, but not before the lugworm has sung of hope and perhaps a better place to exist:

> with its grey and muddy mouth
> Sang that somewhere to north or west or south
> There dwelt a gay, exulting, gentle race
> Under the golden or the silver skies.

Children may be familiar with the nursery rhyme 'There's a Worm at the Bottom of the Garden' and there are many familiar songs that feature worms, often with a theme of worms as food, such as 'Nobody Likes Me (Guess I'll Go Eat Worms)'. By far the most famous may be 'The Worm Song', also known as 'The Hearse Song' or, by its first line, 'The worms crawl in and the worms crawl out (they go in thin . . . and they come out stout)'. As these alternative titles suggest, the recurrence of death and of bodies as food for worms is presented here, and this is a song often sung by children at Halloween.

English comedian Arthur Askey (1900–1982) was known for recording songs about animals, with 'The Bee Song' (1944) perhaps one of his most renowned. Another of his songs on the famous 'His Master's Voice' plum recording label, co-written by Askey, was 'The Worm' (1939). This song is remarkable for several reasons. The lyrics contain some relevant ecological references to things such as mixing soil layers 'leaving little casts in the middle of the lawn', food chains 'filling empty spaces in a blackbird's gizzard' or even recreational activities with the fate of 'sitting on a pin as a small boy's bait'. In addition – and something which

Nobody Likes Me Everybody Hates Me
(I think I'll go eat worms)

No - bo - dy likes me eve - ry - bo - dy hates me I think I'll go eat worms

Long ones, short ones, fat ones skin - ny ones, Ones that squig - gle and squirm

Bite their heads off suck their guts out throw their skins a - way

No - bo - dy knows that I eat worms, three times _ a day

Nobody likes me everybody hates me
I think I'll go eat worms
Long ones, short ones, fat ones skinny ones,
Ones that squiggle and squirm
Bite their heads off suck their guts out throw their skins away
Nobody knows that I eat worms, 3 times a day

If you are feeling unloved and hungry then perhaps this is the song for you?

this book must address at some point – Arthur sings that when he (the worm) meets a gardener's spade, he is cut into two halves and becomes two separate worms.

This brings us back to the perennial question of what happens when you cut a worm in half? Like Arthur Askey in song, many people believe that to cut a worm in two will give rise to two viable worms. It's a question I am invariably asked at each public talk that I give. Equally, when I was involved with research on breeding *Lumbricus terrestris* for soil amelioration, people would suggest that I was wasting my time and could simply keep cutting them

A little-known humorous song, with a few relevant nods to earthworm natural history, recorded by comedian Arthur Askey.

into halves to produce as many as required![23] The fact of the matter is that to cut an earthworm in half and expect to obtain two viable earthworms is a myth. If the cut occurs behind the clitellum, the anterior may survive; in some species that may give rise to a stubby worm with reduced gut length, for example if this happens to *L. terrestris*. In other species, such as *A. longa*, the lost segments may be regenerated from the cut anterior part but can often be seen as obvious regrowth of a lighter colour and perhaps smaller diameter. The severed posterior end of an earthworm will writhe around for some time but, lacking the major nerve and circulatory centres, will ultimately die or hopefully be consumed by a predator as a tasty morsel. Some earthworms have evolved to be autotomic, that is, they have the ability to sever their tail end when attacked. The tail end may act to distract the predator and allow the worm to escape.[24] I have seen some relatively reputable sources suggest that one earthworm can give rise to two by severance, but must re-emphasize that this cannot be the case.

However, with some other types of less complex worms, such as Planarians (flatworms), much more regeneration is possible after deliberate amputation or even maceration into a cellular mixture. Research of this nature has been undertaken for almost a century and a great deal of evolutionary development and genetic biology has been learned, but these are not earthworms.[25]

Few worms have made their mark in the traditional art world, but one painting from medieval Italy shows the emergence of a 'fiery serpent' from the leg of St Roch, a fourteenth-century French pilgrim. Recent research shows that this image, painted in tempera on canvas by an anonymous artist from the fifteenth or sixteenth century, is not a stream of pus exiting an infected wound, as previously suggested by art historians, but a case of Guinea worm (*Dracunculus medinensis*), a type of parasitic nematode that lives in human muscle.[26] The painting of St Roch, with left stocking rolled down, shows the emergence of a 50-centimetre (20 in.) adult worm from his thigh which, as the ancient fiery name suggests, causes extreme burning discomfort. The World Health Organization (WHO) reports that because of the pain there is a tendency for those infected to bathe the exit wound in a water body, causing the adult worm to then discharge hundreds of thousands of larvae. These can then infect water fleas (*Cyclops*), which are the intermediate host and enable completion of the life cycle by reinfection of humans through drinking contaminated water.[27] There is no drug treatment or vaccine to prevent Guinea worm disease, but great progress has been made towards its elimination where it still occurs in sub-Saharan Africa. The number of annual human cases has fallen from 3.5 million in the mid-1980s to 28 in 2018.[28] The treatment towards final elimination of this parasite is provision of clean drinking water through digging boreholes and/or the use of systematic filtering to remove the infected water fleas and break the life cycle. Hopefully there

No cameras available in medieval Italy, but the emergence of a parasitic Guinea worm from the leg of St Roch was recorded in a painting. Anonymous artist, tempera on canvas.

will never again be cause for such a worm to be recorded in a painting or other form of visual reproduction.

The richest earthworm ever recorded was 'Willie', an artistic specimen (species unknown but suspected to be *L. terrestris*) kept by a woman from San Francisco, California. It is reported that his calling was to paint and that during the 1960s he produced nearly two hundred abstract paintings, which reputedly sold for up to £35 each. Willie painted by wiggles after his 'owner' dipped 'him' in the paint, put him on the canvas and allowed his artistic flair to express itself. When one colour was completed, Willie would be dipped into another and allowed to add to and develop the work. It is rumoured that he eventually ran out of inspiration and was returned to the soil.[29] As earthworms obtain oxygen through their skin, it is more likely that this practice would have suffocated the worm(s) concerned and that 'painting by wiggles' were the death throes of numerous Willies who were sacrificed in the name of art. Unfortunately, no examples of the art produced by Willie could be sourced to illustrate his endeavours in this book.

Myka Baum, a graduate from the Royal College of Art (RCA) and now a visual artist, featured earthworms and their activities (burrowing and casting) in photographic works showing that science and art can happily meet. In their show 'The Worm Turns' at the RCA, for example, Baum displayed work that featured, among other aspects of earthworm behaviour, images where *Lumbricus terrestris* had incorporated materials provided into its midden. These included various seeds and even a feather.[30] The latter may be thought of as unnatural, but an ecology text from the 1970s clearly shows that this has been recorded from the field.[31] Darwin also made reference to feathers and states they 'are never gnawed',[32] suggesting an altogether different use than as a food item. The images produced by Baum of *L. terrestris* activity can be considered as stunningly beautiful.

In addition to a carabid beetle and two common weeds of cultivated land, *Lumbricus terrestris* was represented on a postage stamp of the Faroe Islands in 1991 in a series on 'Anthropochora', plants and animals spread by man. The images, drawn by Bárður Jákupsson, show species intimately associated with the farming practices of the islands. A relatively good anatomical depiction of *L. terrestris* is shown on the 6k.50 stamp with farmers tilling the soil in the background. To my knowledge, this is the only illustration of a specified earthworm on a postage stamp from anywhere in the world, although one additional, almost inconsequential, illustration of a 'prehistoric earthworm' appears on a Cuban stamp from 1982, depicted being eaten by an extinct species of

An example of Myka Baum's earthworm-related art (Wormitecture), from *The Worm Turns* (2017).

New World shrew (*Nesophontes micrus*), in a series called 'Animales prehistoricos'.

On a very different tack, as a type of malicious software program, the primary function of a self-replicating computer worm is to infect other computers and remain active on the infected systems. Good cybersecurity, such as firewalls and anti-virus software, can prevent computer worms, but the temptation to click on attractive attachments means that such malware is still a threat and may worm its way on to any system. This is not to be confused in the computing field with the storage technology WORM (Write Once, Read Many).

Faroe Islands postage stamp featuring a lobworm. This was part of a 1991 series featuring organisms dispersed by humans.

Since the 1990s, the movements of earthworms have been studied closely by engineers to devise means of locomotion that require less space, can provide movement over rough terrain or in enclosed spaces, and has potential in several areas of research. The peristaltic movement brought about by the interaction of circular and longitudinal muscles acting upon the hydrostatic skeleton of an earthworm has been used as a model for the construction of earthworm-like robots. These so-called 'soft robots' utilize the same principles exhibited by earthworms – that is, strong environmental adaptability and high deformability. For example, miniature robotics for colonoscopy have been developed for minimal invasive surgery.[33] Electromagnetic drivers can be used to propel a robot the size of a real earthworm (some 12 centimetres long and 1 centimetre in diameter) within the human gut. These robots are designed to reduce patient discomfort and pain during colonoscopy examination and should ultimately lead to reduced death rates from diseases such as colon cancer. As technology is developed, further miniaturization of earthworm-inspired robots is likely to produce micro-catheters capable of navigating further inside the human digestive tract and even the circulatory systems. Away from the human body, earthworm-like robots have also been

A soft robot designed on the hydrostatic skeleton principles of earthworm locomotion.

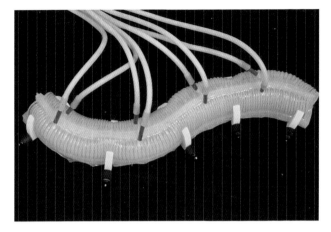

produced to traverse, examine and repair difficult-to-access areas such as underground pipework.[34] These robots are less invasive and therefore less expensive than traditional exploratory methods would be. The flexibility of robots based on earthworm movement is already proving to be of value and will grow in use for exploration of complex spaces.

Further earthworm-inspired robots may soon be put into military service with bio-inspired, soft robot designs that mimic the rhythmic movements of earthworms pushing through soil.[35] The General Electric Company (GE) in the United States was awarded $2.5 million in 2020 through the Defense Advanced Research Projects Agency's (DARPA) Underminer programme to demonstrate the feasibility of a robot that can rapidly and efficiently bore tactical tunnels in support of critical military operations. This research has grown from the expertise of mechanical engineers and computer scientists at Penn State University where, alongside earthworms, elephant trunks and octopuses' arms were used as models of hydrostatic skeletons for soft robotic design.[36]

In 1998 it was reported that Manchester United FC had made a new signing by paying £325 for 5,000 earthworms to help protect the playing surface at their Old Trafford stadium.[37] With a pitch some 106 × 70 metres (116 × 76 yd) in size, this equates to much less than one worm per square yard. Although a sensible addition to soil for many reasons, the scale of this introduction may have made little difference to the aeration and drainage anticipated by such a move. In 2002 an International Symposium

Logo from the 9th International Symposium on Earthworm Ecology (Xalapa, Mexico, 2010). Alec Dempster's design incorporated a local Olmec archaeological head sculpture with multicoloured earthworm representations.

on Earthworm Ecology (ISEE7) was held in Cardiff, Wales, and as part of the programme delegates were afforded a visit to the Millennium Stadium (now known as the Principality Stadium) where they were encouraged to try and collect earthworms from the pitch. The stadium's pitch is based on a palletized system of 7,000 1.2 × 1.2-metre squares and is kept in the stadium for the two playing periods of August to November and January to May, and for other occasions as necessary. Outside these periods the pallets are stored offsite. I am not certain if the management deliberately seeks to keep earthworms out, but using all their expertise more than a hundred earthworm researchers were unable to produce a single earthworm from the turf that made up the pitch at that time. Perhaps we needed help from a charming eight-year-old child and associated grandparents!

Moving forwards, we next explore the ecology and general biology of the remarkable *Lumbricus terrestris*. This might be considered to be everyone's favourite earthworm, but there are a growing number of places where this species is seen as a problem.

5 *Lumbricus terrestris*: (Not Such) a Common Earthworm

If anyone is aware of the scientific name for an earthworm, it is likely to be *Lumbricus terrestris*. This species has been known to science since 1758, named by Linnaeus in the tenth edition of his two-volume *Systema Naturae*.[1] It is a species that will have been studied by many in biology classes at school, as it is given as a classic example of the annelids. *L. terrestris* is referred to as the 'common earthworm', but this is a misnomer, as there are others that are more numerous, and in fact it is a species that demonstrates habits not shown by most earthworms. Over the years scientists have learned much of its basic biology, for example that it is a hermaphrodite with an interesting life cycle and ecology, and that it is one of Darwin's ploughs that feeds and casts at the soil surface, but many questions remain. Does mate choice occur and how? Is mating dangerous? What is an *L. terrestris* midden and what are its full functions? Does parental care occur? These questions and more will be explored here.

L. terrestris is known colloquially as the lobworm, dew worm, nightcrawler, squirrel-tail or twachel.[2] These common names reflect the parts of the world where it is found, its behaviour and use to which it has been put. Dew worm and nightcrawler give an indication of when this animal is most likely to be observed above the soil surface, unlike other earthworm species that only appear after rain. Europeans out walking in the early morning may well

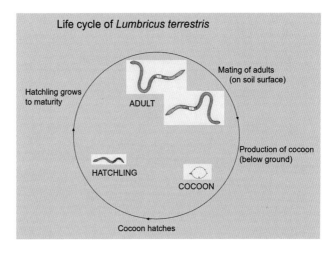

Life cycle of *Lumbricus terrestris*

Mating of adults
(on soil surface)

Hatchling grows
to maturity

ADULT

Production of cocoon
(below ground)

HATCHLING

COCOON

Cocoon hatches

observe *L. terrestris* as the dew worm, still working to gather food at the soil surface in the glistening grass just after dawn. The name 'nightcrawler' comes from the same origin, as these animals explore the soil surface close to their burrows from dusk until dawn, usually seeking food such as fallen leaves, but possibly seeking a mate. If you're an angler looking for a specific bait, this may be a good time to search for them without recourse to digging. As far back as the seventeenth century it was noted by Izaak Walton in his book *The Compleat Angler* that it is 'the twachel, or lob-worm, which of all others is the most excellent bait for a salmon'.[3]

The name 'squirrel-tail' comes from the flattened rear end of an otherwise cylindrical earthworm. Most authors have written that this physical feature has evolved to allow the animal to anchor itself into its burrow when feeding, allowing it to retreat rapidly if a predator should be sensed, or at least not be pulled out immediately. This is no doubt part of the reason for this adaptation, but research using infrared filming in darkness has shown that on occasion this earthworm will venture beyond the safety of the entrance and completely exit from its burrow when feeding. Although not always successful, it was recorded that some risk-taking *L. terrestris* can use the posterior of their body (the flattened tail) as a sense organ and, by using a side-to-side motion in contact with the soil surface, they feel their way back to the burrow

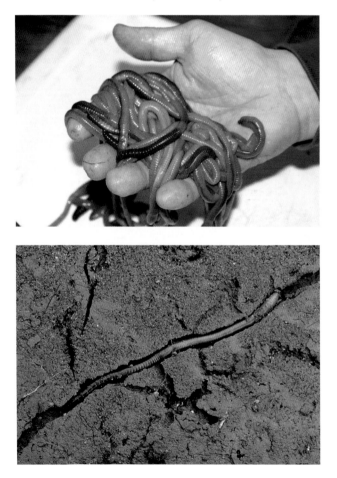

A handful of lobworms, their adult status determined by the presence of an orange clitellum. Average individual biomass is around 5 grams.

An adult lob worm clearly seen within its burrow, an unusual horizontal construction, exposed beneath a paving slab.

entrance.[4] During this time the worm's head appears totally inactive, as if the tail has taken control of the body and sensory pits on the skin are tracking the mucus trail that it has left on the soil surface. This homing behaviour, if successful, extends the feeding range of the individual, but must be traded against a failure to return. It is certainly an area of behavioural ecology worthy of further research.

The dew worm is the largest earthworm found in Britain and across most of Europe (excluding a close relative, *L. badensis*, plus a few species endemic to the Carpathian Mountains and from Southern France). It has a Eurasian origin, but due to transportation by humans is now found in agricultural and forest ecosystems in North America. It is widely distributed across numerous terrestrial habitats but usually most numerous in grassland, including parks and gardens. It is found where tree species and hence litter fall may influence soil properties, although less common in woodland. It prefers a soil pH in the alkaline range, up to around 10, but can exist in the slightly acidic conditions found in some coniferous woodlands. This species is particularly well represented in clay soils; adult worm abundance does not normally exceed about thirty per square metre.

L. terrestris lives within a permanent, near-vertical burrow that can extend more than a metre into the soil. The cylindrical burrow generally has a diameter just larger than that of the resident earthworm, usually around 10 millimetres ($^2/_5$ in.), through which it can move up and down using muscular contractions and the setae on its body. But why would the dew worm create such a lengthy burrow into the soil, when the distance far exceeds the length of the animal (perhaps only 30 centimetres (12 in.) even when fully extended)? An obvious answer might be to escape from potential predators such as birds, and indeed after detection of vibrations from the soil surface, *L. terrestris* will rapidly withdraw into its burrow. Nevertheless, some mammals such as European badgers (*Meles meles*) can 'suck' resting earthworms out of their burrows from just below the soil surface, and so even a deep burrow may not be helpful here. Of note is that under ideal conditions, badgers have been recorded to consume between 130 and 200 mature *L. terrestris* from pasture in a single evening.[5] The deep burrow must therefore have other functions. Before going further, it is useful to more fully describe the context of the burrow, as it really forms a part of what has been termed the 'burrow-midden-complex'.[6] The burrow of the dew worm does not just end as a hole at the soil surface. The entrance is carefully managed by the resident earthworm and is covered and surrounded with a collection of material gathered (by mouth) from the locality. This material may consist of twigs and leaves, if in a woodland, but could also be a collection of non-biological items such as stones (ever thought how those small cairns on your driveway were formed?), which then throws up even more questions. Is this a food source? If not, then what is the function?

In grassland close to trees, protruding leaf stalks show the position of lobworm middens. The leaves have been drawn into the burrow below for consumption.

The midden is formed at and around the burrow entrance and is an integral part of the life of the animal that creates it. In addition to the collected item from the vicinity of the burrow, the earthworm casts upon the materials to 'cement' them together. Filming has also shown that after a night of activity, *L. terrestris* may also produce casts to form a temporary seal at the burrow entrance in a swirled ice cream-like fashion.[7] This is no protection from predators but may act as a way of stopping moisture loss. It is of interest that *L. terrestris* creates a burrow within which, because of its body form, it is able to completely turn around so that it can be 'head-up' to enable feeding, but also 'tail-up' for casting, as desired. The burrow-midden-complex may also be considered as an extended phenotype of this earthworm. That is, the physical structure of the burrow and midden, just like a web for a spider and a nest for a bird, should be included in the effects that the genes of the organism can have on the environment. This is an idea that has been further explored by other authors but is beyond the scope of this book.[8] The presence of *L. terrestris* in any habitat can be determined by careful observation of the soil surface for the presence of middens. These can usually be seen quite easily, but in some pastures feeling the surface of the grass may assist in location. Confirmation that a midden has been located can be made by looking below the collection of material or slight mound for a distinctive burrow opening.

Although he did not name this species in his book, Darwin undoubtedly recorded the activities of *L. terrestris*. His investigations of leaves taken into burrows must have been by dew worms and he also speculated on the reasons for their collection and determined quite rightly that most would have acted as a source of nutrition, either directly or indirectly. Darwin was also convinced that 'burrow plugging', as he called it, was not an effective deterrent 'to conceal the burrows from scolopendras' (predatory

After midden removal, an adult dew worm is enticed from its burrow with a vermifuge. Smaller worms nearby had been present within the midden.

beetles).[9] My own unpublished findings have shown that when seeking to extract dew worms from their burrows, a surprise can be the appearance of a larval *Pterostichus* beetle instead. No doubt the previous resident, the worm, had become a meal for the latter, something we know to occur, as revealed from DNA analysis of the gut contents of these beetle larvae.[10] A more likely suggestion for the presence of a midden, and seen by Darwin as the most probable explanation, was that 'the mouths of the burrows are closed in order that the air within them may be kept thoroughly damp.'[11]

It is in spring and autumn that *L. terrestris* is most active in temperate regions, because soil moisture and temperature conditions are normally ideal. Field and laboratory experiments have shown that this animal has optimal ranges in which it exists and extremes beyond which it cannot survive: for temperature this is a lower limit of 0°C, at which tissues freeze, and an upper limit of 28°C, at which enzymes are denatured. It is thought that the depth of a burrow may indeed be determined by soil moisture conditions. When creating somewhere to live, it is believed that

L. terrestris burrows down until it contacts moist soil. This then means that the burrow can draw up a continuous flow of moisture from depth. The capping of the burrow with a midden then makes some form of sense, as suggested by Darwin, as it prevents undue water loss through evaporation. It should be noted that the resident earthworm lines its burrow with a mucus secretion from its body, which aids vertical earthworm movement and prevents not only abrasion against the soil, but moisture from being lost into the layers of soil surrounding the burrow. The mucus lining itself, which needs to remain moist, is a very active area in the soil and regarded as a 'hotspot' for microbial activity. The burrow-midden-complex accordingly has at least one major function, keeping the dew worm moist so that it can function to move vertically and feed at the surface.

One technique for examining the extent of a *L. terrestris* burrow is to fill it with a substance that flows down the full length and then solidifies to permit the volume of the burrow void to be

Middens in reduced tillage agro-ecosystems are straw-rich, gathered by the resident lobworm from the immediate vicinity.

measured, and even extracted as a unit by digging it out intact. This was a technique we examined a while ago. After trying such materials as molten lead (not to be advised on health and safety grounds), use of a polyurethane resin mixture, made more visible by colouring with paint, was found to work extremely well.[12] The resident earthworm needs to be extracted first (using a vermifuge), the burrow allowed to dry and then the resin poured in. This then sets within minutes and is left to totally harden over 24 hours. An area close to the burrow can then be excavated with a large digger and the burrow exposed laterally by standing within the excavated hole. This permits the full extent to be seen *in situ*. Should there be any unexpected shapes – for example, caused by the presence of large stones that the worm has been unable to burrow through – then such deviations can be easily seen. Equally, any specifically created side burrows can be recorded. Extensive work of this type in the UK, USA and Finland has suggested that burrows of dew worms are generally vertical but may have a slight shallow incline near the surface where the worm may lie at rest without a need to expend energy by defying gravity using muscular contractions and setting setae against the burrow wall. Excavations also showed no evidence of a chamber at the base of the burrow, as suggested by some authors.[13] It is apparent that some basic information about the biology of this species is still to be determined.

Research on the burrows of *L. terrestris* in agricultural fields where subdrains are present has found that the position of these subdrains greatly influences the abundance of earthworms nearby. This is because soil moisture content over the course of an annual cycle can be vital, with excess water as much of a problem as insufficient water supply. More dew worms were found where drains were present than at locations in between the drain lines.[14] Further investigations using resin-casting of burrows showed that there was a direct interaction between these burrows and the subdrains in the fields.[15] It appeared that the earthworm burrows were orientated towards the subdrains (at a depth of around 1 metre) and even dug deliberately by the worm to meet them. This most likely assists the resident worm to have a way of utilizing the water removal facility from the heavy clay content soils. Here we find a mutually beneficial interaction between an ecosystem engineer (the earthworm) and drainage engineers (ourselves.

In addition to temperate regions, dew worms can be found in sub-boreal zones as far north as the edge of the Arctic Circle.[16] At such latitudes the soil regularly freezes from the surface down to 50 centimetres (20 in.) and at times a greater depth than this. Therefore, another good reason for *L. terrestris* to have a deep burrow is to be able to escape from freezing soils in which adult worms would most likely perish, although some other earthworm species are known to possess a natural anti-freeze.[17] Research during winter investigations, using large excavator machines to break through frozen topsoil, revealed that *L. terrestris* was inactive, head up, at the base of its burrow awaiting a thaw and the opportunity to forage once again.[18] In addition, dew worms have been shown to be active in the midsummer months in Finland, when the sun barely sets. This is the time of 'white nights', so dew worms cannot exit their burrows under cover of darkness and therefore need to override their natural instinct to shy away from light. The number of months for active foraging at higher latitudes is few, so feeding is required when possible. Experimental work undertaken to compare Finnish animals with the same species collected from lower latitudes (UK and USA) and then taken to Finland, showed the native animals to be much more active in the summer conditions than their cousins from afar.[19] This almost certainly relates to a need to build up bodily food reserves and collect organic matter into the burrow to feed upon when the

A vertical (1-metre-deep) *Lumbricus terrestris* burrow, filled with white-coloured resin and exposed through digging, in a drained agro-ecosystem.

Removal of a frozen layer from the soil surface in preparation for defrosting and examination for over-wintering earthworms and cocoons. Adult *L. terrestris* were found inactive, at a depth of 1 metre.

upper parts may be frozen during a prolonged winter. Under controlled light conditions (twelve hours light: twelve hours dark), *L. terrestris* has been shown to become active within the first hour after darkness and then feed and mate throughout the dark period with a decline in activity as the hours progress.[20]

It has already been mentioned that *L. terrestris* is normally active at the soil surface under cover of darkness (excluding activity during 'white nights' and extended mating). Although it has no eyes, like most earthworms, light sensory cells are present, particularly in the most anterior segments, and worms will retreat rapidly when brightly illuminated. In addition, chemosensitivity exists, which on a gross scale allows reaction to a range of pH levels but can be much more subtle.[21] The selection of nitrogen or sugar-rich leaf litter by *L. terrestris* is possible owing to the worm's sensitivity particularly in the first segment (prostomium) where as many as seven hundred taste receptors are present per square millimetre.[22] Perhaps even more important to *L. terrestris* is a sensitivity to vibrations. Anyone who has tried to collect dew worms feeding at the soil surface will have witnessed a rapid retreat into the burrow following an accidental heavy footfall. The setae and nerve endings are extremely sensitive, demonstrating that foraging badgers in pasture must be very delicate feeders. Darwin demonstrated that earthworms (*L. terrestris*) possess no sense of hearing: when he made deliberate noise such as the 'shrill notes from a metal whistle' or 'the deepest and loudest tones of a bassoon', no earthworm activity was recorded.[23] However, although indifferent to sounds we perceive, Darwin showed their extreme sensitivity to vibrations by placing earthworms in a pot of soil on a piano and striking various notes. The earthworms rapidly withdrew into their burrows. When the same notes were played but the pot of earthworms was not placed on top of the piano, the earthworms didn't react, indicating that it wasn't the sound, but the vibrations, that they were responding to.

A hermaphrodite, the dew worm has both male and female reproductive organs. It can therefore operate as both: when it mates it will simultaneously produce sperm to fertilize its partner and be fertilized itself. Unlike some earthworm species, it is not parthenogenetic and so still requires cross-fertilization with another individual, making it amphimictic. As *L. terrestris* lives in a permanent vertical burrow, there is no opportunity for a chance meeting with another adult worm within the soil. Therefore, to find a mate there is a need to explore and search on the soil surface. To avoid most diurnal predators, mating attempts take place at night. In darkness, as with feeding, the dew worm ventures on to the surface of the soil and, keeping its tail in the burrow, can explore a circular area with a radius equal to the length of the outstretched worm. Mating is possible with another *L. terrestris*

A recreation of Darwin's experiment on earthworm reaction to the playing of a bassoon.

encountered doing the same. It is as well that burrows are found close to each other or this would not be possible. This encounter, however, is not by chance, but likely due to settlement of dispersing individuals in areas where conspecifics are detected by mucus sensed on the soil surface and possibly by discovering middens. It is remarkable how far mates are prepared to stretch to achieve their objective. Heading to work one morning, I observed a worm in my garden mating with a worm below the boundary fence to my neighbour's garden. Both were fully stretched out. The chance of such an encounter of two worms at the surface may be minimal, so a behaviour has evolved to increase the mating possibility, and this is for one potential partner to partially enter the burrow of another to initiate a mating.

When two adults meet at the surface, or one partially enters the burrow of the other to begin a mating attempt, there is a period of interaction that was unknown until recently. A two-way 'checking out' occurs before any mating happens. This involves reciprocal burrow visits by the pair, usually over a period of some minutes with one following the other back to the potential mate's burrow and then vice versa. This may occur only once before mating, but has been recorded to happen thirty times over a period of up to two hours.[24] But why? There is obviously some selection pressure here, and in some way each potential partner is assessing the fitness of the other. Is it to assess the burrow (and therefore resident earthworm) dimension, volume/quality of organic matter (in the midden) or some unseen chemical cue? Experiments

Widely elongated worms stretching under a fence between two gardens in order to mate: the lengths that worms go to to achieve reproduction.

were undertaken to investigate one of these possibilities, following a size-related hypothesis. Adult *L. terrestris* were collected from the field and subdivided into either 'small' or 'large' by mass. In groups of three, these were then allowed to burrow into a container of soil, with plenty of room for separate vertical burrows. Each group of three comprised either two small and one large, or two large and one small. Over time, video recordings were made of meetings between the trios and how they behaved towards each other with respect to mating. It was found from numerous observations that small adults chose to mate with other small adults and large adults with other large adults.[25] There had to be a good reason for this.

Mating of *L. terrestris* can be a prolonged and quite forceful activity, with the whole copulatory process from meeting, burrow visits and separation taking more than three hours.[26] It takes place on the soil surface with the tail of each worm in its own burrow. The two animals having 'checked each other out' then enter a

Mating *Lumbricus terrestris* within an experimental set-up. Note that the tail of each remains in its own burrow.

ritualized behaviour whereby they need to lie alongside each other in a very specific position to line up necessary segments (numbers 5–15) to effect exchange of sperm (recall that each is acting as a male to the female part of the other). They possess no penis, so sperm needs to flow from male pores to a sperm receptacle by way of rhythmic contractions of the body segments. For this to occur and be successful, the alignment needs to be exact and the two worms need to be tightly bound to each other. This is where some specialized structures on the body come into play. These are copulatory setae (forty or so in number) and used during the lining-up phase of copulation to pierce the skin of the partner and lock them together in exactly the correct position for sperm transfer. They also inject substances from glands into the partner that influence sperm uptake.[27] Once alignment is achieved, sperm can be pumped from each worm to the other along sperm grooves between the sides of their firmly linked bodies.

After mutual sperm exchange, it is time for the two earthworms to separate. This involves a movement, by each, to fully return to the home burrow and disengage from the embrace of the copulatory setae. This can be quite a wrench, as it requires each partner to pull against the other and with great force rip out the piercings. If the partners are of a similar size, this usually results in a clean separation, with each mated worm returning successfully (backwards) to the home burrow. However, in cases where unequal-sized individuals mate (large with small), the larger individual often pulls the tail of the smaller *L. terrestris* from its burrow before separation, particularly if the burrows are not close together. This is a catastrophe for the smaller earthworm as it will be unlikely to easily regain its home burrow and may become predated at the soil surface. It is, however, also a disaster for the larger partner, as it will have lost the potential to act as a 'father' to the offspring of the smaller worm. Therefore, in these

Diagrammatic representation of the dew worm mating sequence. After a series of reciprocated burrow visits, the two hermaphroditic adults copulate on the soil surface and exchange sperm over a period of more than two and a half hours (adapted from the *Journal of Zoology* (1997)).

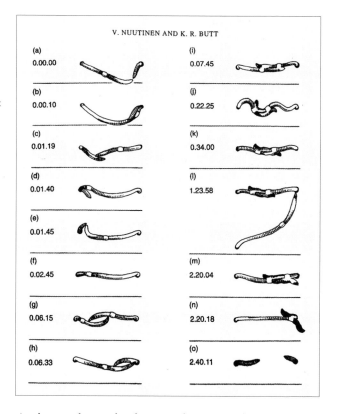

V. NUUTINEN AND K. R. BUTT

(a) 0.00.00

(b) 0.00.10

(c) 0.01.19

(d) 0.01.40

(e) 0.01.45

(f) 0.02.45

(g) 0.06.15

(h) 0.06.33

(i) 0.07.45

(j) 0.22.25

(k) 0.34.00

(l) 1.23.58

(m) 2.20.04

(n) 2.20.18

(o) 2.40.11

simultaneous hermaphrodites it is advantageous for both partners to be successful in mating, acting as a male and as a female. This, it is thought, is why the pre-copulatory behaviour of 'burrow-visiting' occurs, to gauge the size of a potential partner (size-assorted selection) to try and ensure a successful mating, acting as both male and female. It can be seen here that the squirrel-tail shape of the animal is of vital importance, since it grips the inside of the burrow, not to mention any sensory role that it might also play.

It should be noted that the act of mating is another behaviour that overrides the dew worm's negative reaction to light. In an experimental situation of twelve hours light and then twelve hours where surface behaviour was recorded in darkness under infrared light conditions, most matings were initiated soon after darkness fell and so were completed before lights came on with an average length of copulation lasting 2.5 hours. However, in a small number of cases, matings were begun late in the dark period so that the two worms were still joined when the lights came on. This did not deter the copulation, which lasted as normal and afforded the opportunity for clear photographs to be taken. In addition, this has been seen under more natural conditions, as mentioned above with my own and my neighbour's garden-based *L. terrestris*.

As detailed, sperm is stored, and mated individuals have been shown to produce cocoons for up to twelve months after the mating. A study on the hatchability of cocoons found a range of 62 to 76 per cent hatched over the five months following a single mating of virgin animals, which decreased to about 11 per cent in the sixth month, and cocoons produced after that failed to hatch. Median total production of viable cocoons was five per individual, with a range of 0–21. From a single mating, sperm may be stored for as long as eight months.[28] Mating under natural conditions is not limited to a single occurrence and under experimental conditions matings may happen every third night, often with different nearby partners. This can lead to sperm competition (within the sperm store) and enhanced fitness of offspring. As Gilbert White of Selborne noted in 1777, these animals 'are much addicted to venery', an archaic term for sexual indulgence.[29]

Having mated, the adult *L. terrestris* can then produce cocoons. This takes place below the soil surface where the adult creates a specific, perpendicular side branch from its deep burrow that is only about 3–5 centimetres in length (1–2 in.). This is where a single cocoon is deposited. We know this from filling burrows using polyurethane resin which we subsequently extract, and from observations in what are effectively two-dimensional, glass-sided wormeries, also known as 'Evans' boxes'.[30] The cocoon is formed by secretion of proteinaceous material from the clitellum of the adult, producing a cylinder of tissue. The animal then locates its anterior (head) in the position where it wants to deposit the cocoon (in the side burrow), and by rhythmic muscle movements the cylinder is moved forwards. During this passage the adult deposits its own eggs and nutrients plus stored sperm into the cylinder. When the cylinder slips off the head of the worm, the ends close together to form a lemon-shaped cocoon, which is about 7 millimetres in length and olive green in colour. The adult is then known to deposit soil and organic matter (leaf mulch) within the side burrow, to effectively seal off the cocoon. There is debate as to the function of this material, but it might act as a first meal for the juvenile worm once it hatches or could be to deter any predators that enter the main burrow from finding the

Lobworm cocoon produced in a side (cocoon) burrow of its parent and surrounded by organic material deposited by the earthworm. An example of parental care?

cocoon, effectively an encapsulated meal. Cocoon development is temperature-dependent, so the time of year when the deposition occurs and latitude at which the earthworm is found will determine the time taken to hatch and may account for the chosen depth below the soil surface.

Cocoons tend to be deposited relatively close to the soil surface (within 10–15 centimetres/4–6 in.), so can benefit from soil warming in spring.[31] Freezing of soil around the cocoons, as occurs in some countries, has been shown to have no adverse effect, and cocoons hatch successfully after the soil thaws.[32] Laboratory work has shown that cocoons take 270 days to develop at 5°C, but only 180 days at 10°C and 70 at 20°C (41, 50 and 68°F, respectively).[33] Ninety-nine per cent of *L. terrestris* cocoons produce a single hatchling weighing around 50 milligrams (1/500 oz). If twins are produced, they are proportionately smaller and the same applies for triplets, which I have witnessed only once. The hatching success rate, from laboratory experiments with freely mating adults, is around 70 per cent.

One unknown in the life history of *L. terrestris* is what happens after a hatchling emerges. It is believed that it may feast on the leaf material provided in the side branch of the adult burrow, but it must then make its way to the soil surface, because it would be

Cocoons of *L. terrestris* in a Petri dish of water in which they will develop and hatch over a period of months.

Cocoons of *L. badensis*, a large species from the Black Forest in Germany. Cocoon size is a function of adult earthworm size.

incapable of burrowing through the soil at such a small size. Does it simply crawl up the inside wall of the adult burrow using the mucus lining to assist or does perhaps the adult play a part (as crocodile mothers do when their offspring emerge from the nest and are helped to the water)? We have yet to witness such behaviours and this is a research field that is still wide open. We do know that hatchling *L. terrestris* (and other small worm species) can be found inhabiting the soil of the midden at the burrow entrance.[34]

The materials of the midden, gathered from a circular surrounding area – to a worms-length radius (tail always in burrow)

Newly emerged hatchling dew worm beside cocoons. Mean mass at hatching is around 50 mg.

– are cemented together with castings (faeces), forming a very fertile area. The hatchling/juvenile *L. terrestris* may use the midden of the parental worm as a form of nursery, staying there for a time and growing a little by making use of organic matter enriched by microorganisms. Laboratory research has shown that this species does not construct a vertical burrow of its own until a mass of around 1 gram (4/100 oz) has been reached. It is likely that in the field, juveniles disperse from the parental midden, when environmental conditions permit, under cover of darkness when it is wet and not too hot or cold, and rainy spring and autumn nights may be ideal. These bouts of dispersal may be driven by the earthworm perceiving rainfall on the soil surface (as previously mentioned with respect to foot-trembling by birds). Growth to maturity, reaching the clitellate condition, in the field may take up to a year in Britain but longer in northern Europe. Under laboratory conditions this time can be halved through provision of optimal conditions of temperature and quality food.[35]

Dispersal of adults has been recorded to reach around 20 metres (65 ft) in a single night, with the animals settling either when a favourable spot is found near to others or when daylight is approaching.[36] A further intriguing aspect of *L. terrestris* behaviour is that burrows may be inherited by offspring or reused by others. A study in a Finnish woodland indicated that the fidelity of burrow positions over a sixteen-year period was quite true; as this time span exceeds that of the predicted life of *L. terrestris*, reuse was likely to have occurred.[37] Given the way that hatchlings must emerge from the adult burrow, it is possible that a younger worm could adapt a vacant burrow (which has been abandoned because of either the death or dispersal of its previous owner) to itself, by narrowing the diameter through internal casting until it grows to 'fill' the burrow. Such occurrences have been observed in laboratory settings, where adults have been removed from

Evans' boxes after cocoons have been produced and the hatchlings left *in situ*.[38] It has already been stated that the greatest abundance of *L. terrestris* usually found in the field is around thirty adults per square metre. Experimental work within 'earthworm-fenced' square metre field plots demonstrated why this may be so. An earthworm fence is a sheet of plastic that extends below the soil for some 20 centimetres (8 in.) and above the soil to the same height, so preventing *L. terrestris* from escaping except into 'tunnel traps', openings leading to containers set along the base of the fences to capture any worms trying to leave the plot.[39] Part of the experiment looked at the effects of adding extra adults to the area (artificial colonization to reach superabundance) and many were found captured in the traps. We believe that this maximum abundance is a function of two competing elements in the lives of the dew worm: sex and food. Burrow location needs to be close enough to a neighbour to facilitate the over-surface reaching and mating behaviours, but also needs to be sufficiently distanced to reduce competition for food (fallen leaf material primarily) and midden building materials. We therefore found that on average each burrow is about 13 centimetres (5 in.) from the nearest neighbours and a relatively regular pattern of burrow spacing is usually found.[40] The regularity increased with a greater burrow number, revealed by the presence of middens.

Where organic matter is not deposited evenly in a location containing *L. terrestris*, initially the middens of those with a superabundance of material will be supersized. However, this situation does not last and what follows is a 'relay' of straw movement, where those close to the middens with large masses of straw take material to their own middens, part of which, in turn, is moved to adjacent smaller middens and so on. In this way we believe that organic material at the soil surface may be redistributed widely in a biologically active way, alongside the action of wind and rain.

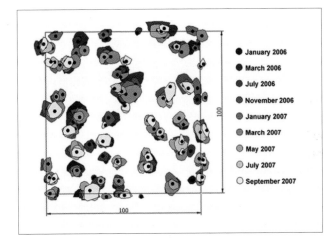

This organic relay movement of material was clearly seen from investigations in an agro-ecosystem but has probably gone unnoticed in most natural ecosystems, such as grassland and woodland where *L. terrestris* is present.[41] It is certainly something worthy of further investigation as it may have a major long-term effect on soil fertility.

Unlike many invertebrate species that grow to a given size, earthworms have the capacity to continue growing if optimal environmental conditions prevail and predation is avoided. Under laboratory conditions of constant temperature, excess good-quality food material and without intraspecific competition, *L. terrestris* were able to grow to a size of 10 grams (7/20 oz) within fifteen months.[42] This is twice the mass that might be expected from field-collected animals. Nevertheless, it is thought that where similar optimal, predator-free conditions occur in a natural landscape, a similar large size can be attained. For example, in deciduous woodland at Papadil, a relatively predator-free, remote location on the Isle of Rum in Hebridean Scotland, *L. terrestris* with a mean mass of 11.7 grams (²/₅ oz) were located in 2006, making them the largest then recorded in Britain.[43] An animal of 15 grams (¹/₂ oz) was reported in Finland in 2011 and then in 2016 a specimen of 26 grams (1 oz) was found by a gardener in Widnes, Cheshire, and taken to the Natural History Museum in London. This exceptionally large *L. terrestris* (known colloquially as 'Dave') is now the record holder for largest size, which has been attributed to its life before capture within a vegetable-rich area of a domestic garden. The natural lifespan of *L. terrestris* is uncertain, though known individuals have lived for six years in captivity in our laboratory. We believe that those from the Isle of Rum may well have been more than a decade old. There is plenty of scope here for investigations of earthworm longevity and maximum size.

It is many decades since anyone examined in detail the bait market for *L. terrestris*. These animals are generally regarded as a superior fishing bait over smaller, potentially fetid-smelling compost worms. Recent social scientific research has documented how agricultural systems and a willing labour force are organized to supply the world market for the 'Canadian Nightcrawler'. Based on 59 semi-structured interviews, the findings showed that an

Almost three times the average size, *L. terrestris* at a remote location on the Isle of Rum may live longer due to little disturbance, a good food supply and limited pressure from predation.

estimated 500 to 700 million earthworms are picked annually from farm fields between Toronto and Windsor, Ontario.[44] Some dairy farms emerged as effectively *L. terrestris* production sites because of their perennial alfalfa crops, heavy manure application and reduced tillage practices. Due to this, *L. terrestris* has become the most lucrative crop in the region with many farmers leasing land to worm-picking operations for more than CAN$1,000 per year (U.S.$750 at 2020 rates), some four times the regional rental rates. Worm-pickers have historically been recent immigrants to Ontario but can earn CAN$20 per thousand worms and can pick more than 20,000 worms per night at the soil surface in optimal field conditions (moisture, temperature, wind and moonlight). Payments tend to reward speed and efficacy. This earthworm-focused arrangement between dairy farmers, soils and worm pickers opens up even more opportunities for further research, but here by combining socio-economic, agronomic and ecological elements of *L. terrestris* harvesting.

So far we have seen *L. terrestris* acting in many positive ways, providing services to local ecosystems. However, in some areas of the USA and beyond it is viewed as a problem species. To explain this requires a little background information. *L. terrestris* was accidentally introduced to the North American continent as far back as the time of European settlement in the sixteenth or seventeenth century, most likely in soil associated with trees. More recently, discarded fishing bait of this and other earthworm species has seen further spread into natural ecosystems. *L. terrestris* is therefore an introduced species to the USA and Canada via the agency of humans.[45] Now considered an essential part of some agricultural systems, its ecological engineering role in the soil by burial of leaf litter, bioturbation and releasing nutrients is alien to some ecosystems. Its invasion fronts have introduced this species throughout Canada and northern hardwood forests of the United States.

L. terrestris therefore alters ecosystems, and the long-term trajectory of previously earthworm-free forest communities may be drastically changed by organic matter (leaf litter) removal from the soil surface layers, nutrient cycling and availability, seed collection and burial, and even a change in microorganism balance through direct consumption or changes in soil conditions.[46] These problems are not confined to North America and research from the early twenty-first century in Europe has shown that *L. terrestris* has invaded new habitats in Romania, replacing mostly endemic or rare species of the former earthworm communities that were recorded over previous decades.[47] The dew worm may be very competitive and a successful ecosystem engineer, but is not always a desired species in an ecosystem.

The final chapter of this book will return to earthworms in general and continue to examine expanding and future directions of research. Our own lives and those of earthworms may be intimately linked and we would do well to ensure that they and the soils in which they are found are not neglected or damaged.

6 The Future of Worms and Humans

We all owe a great deal to earthworms, as they were instrumental in soil formation and are largely responsible for the natural maintenance of the earth from where we derive our food plants. Their ecosystem services are only disrupted or compromised by humans, for example through modern farming practices that use inversion ploughing, artificial fertilizers and pesticides, but it is possible to reverse and rectify such disruptions and return to more natural (earthworm) ploughing. The coming decades are likely to see the implementation of less-intensive farming methods as a result of the greater realization of the planet's finite resources, which are vital not only to us but to earthworms, although this movement may be compromised by a growing human population. This chapter seeks to project forwards and, at differing scales, show how earthworms may impact upon us in the future. It will consider active areas of research and focus on familiar topics, highlighting how harnessing the work of earthworms could be even more beneficial to us. Finally, it will look at some of the bigger issues with which we are faced and see how earthworms may impact upon these and are themselves impacted upon by our activities.

Much has been learned about earthworms since the times of White and Darwin, and earthworm-related research moves forward at a great pace. Anyone who now begins to study earthworms has a tremendous task to obtain a full grasp of the required literature. A comparison drawn between 1880 and 1920 (the years following the publication of Darwin's book) and 1980 to 2020 shows that the volume of publications is more than 160 times greater, with an exponential growth of research on earthworms over recent decades. On average, two scientific papers were published every day in quality journals between 2010 and 2020.[1] This reflects a growing understanding of the major effects that these organisms have in soils, how they can be used as ecological tools for processing organic waste materials and used as model organisms in the field of ecotoxicology.

From a practical perspective, future earthworm investigations need techniques to be developed that allow a more intimate 'worm's-eye-view' of the world. Currently, to study earthworms in the field we must disrupt or invade their subterranean world. To avoid this and obtain a more realistic view, use has been made of rhizotrons, large glass-sided, below-ground observation windows, but these are still only a two-dimensional exposure of a three-dimensional space.[2] It would be useful if we could enter that space without interfering with it. A colleague from Finland obtained a second-hand medical endoscope for examination of the human oesophagus, and inserted this, with inbuilt camera, down into a *Lumbricus terrestris* burrow. The footage obtained of the worm coming up to meet the instrument and seemingly thrusting towards it suggested the way that a resident *L. terrestris* might react to a visit from a potential partner or to a potential burrow competitor. Such an observation was novel and revealing, and suggested that production of miniaturized cameras and other small tracking instruments might still uncover much of the basic behaviour of earthworms that can currently only be hypothesized.

Following the fate of individual earthworms may prove to be valuable in population studies relating to survival, distances moved and other behaviours. With larger organisms, such as birds, the attachment of leg rings has led to a wealth of data collection that has revolutionized our understanding within ornithological study.[3] For earthworms, however, the attachment of an external marker to a mucus-covered, transformable-bodied animal is close to impossible as it would disrupt the natural movements of the worm. But what of an internal tag? Experimentation using a coloured, injectable biopolymer, which can be seen through the earthworm skin and is able to move with the shape changes of the animal, has proven to be partially successful. Use of Visible Implant Elastomer (VIE) has shown that retention of the tags in laboratory studies can last for more than two years without any impact on earthworm survival or reproduction. After this time it became more difficult to see the tag as the earthworm produced tissue internally that began to cloak it.[4] VIE tagging has also been used successfully in the field and allowed assessment of *L. terrestris* population dynamics at the scale of a square metre.[5] Addition of tagged animals to an already fully populated area

A futuristic glimpse of earthworm science enabling a worm's-eye view of soil activity without disturbance of the soil.

was undertaken to establish if 'supersaturation' could occur by provision of extra food or whether the known effects of spatial requirements for feeding and mating might override this. VIE-tagged animals, captured on leaving the experimental areas, were easily distinguished from resident dew worms that were tag-free. This method has yet to be exploited as fully as it might, due to a perceived difficulty in administering the tags, which becomes easier with experience. In addition, the tags only last for the life of the animal, so perhaps this technique is only of value in more longer-lived earthworms or for use in relatively short-term experiments. In the laboratory, VIE tagging has been put to good effect in ecotoxicological experiments with earthworms where a gradient of contamination was set up and individuals with different-coloured tags were added at specific positions along that gradient.[6] After a short period of time, such as seven days, it is then possible to determine which earthworms have moved to which positions, most likely in response to the level of contamination in the soil, and whether point of introduction was an important factor.

Allolobophora chlorotica marked with a Viable Implant Elastomer (VIE) tag (seen as a white patch). Tagging can identify individuals and assist in population investigations.

Experimental investigations of earthworm behavioural responses to soil contaminants in ecotoxicology gradients.

The use of earthworms as bioindicators of potential soil pollution is a developing area of research. Due to the intimate contact of their moist skin with soil (external effects), direct consumption of soil (internal effects) and general sensitivity to toxins, earthworms have become a standard test organism.[7] Such tests examine the effects of potential contaminants on key life-cycle elements, including survival and reproduction. However, these standardized tests, developed over past decades, specify use of the epigeic (compost) worm *Eisenia fetida* as the focal earthworm, most likely because it is easily obtained from suppliers. Colleagues and I have argued for many years that this species is inappropriate as it does not really live within the mineral soil, and current thinking is now moving towards use of endogeic species such as *Aporrectodea caliginosa*.[8] This is a widespread species and more likely to give test results that are of value with respect to potential contaminants in soils and their effects on soil fauna. If we are seeking to use a 'canary in a cage' approach, as was once the way to detect noxious gases in coal mines, let's at least ensure that we have the right type of canary to start with. Such ecotoxicological results are often built into models, but these are not confined to contaminated soils and can be as important in agro-ecosystems.

In terms of modelling the risks of plant protection products (pesticides) towards earthworms, a similar approach has been taken whereby the 'wrong' ecological types of earthworms were considered. The models have therefore tended to be relatively simplistic and resulted in some uncertainty when applying the results to agro-ecosystems. More recently, environmental risk-assessment modelling across Europe has offered a powerful tool to integrate the effects observed from laboratory studies with the environmental conditions under which field exposure of populations is expected. Here though, all ecological groupings of earthworm (epigeic, endogeic and anecic) were considered and the anecics were even subdivided. This was to recognize that *Lumbricus terrestris* is a very different animal in terms of its behaviours to other deep-burrowing species such as *Aporrectodea longa*. This consideration demonstrates that modellers are beginning to have meaningful discussions with soil ecologists and an appreciation of the types of earthworms present in agro-ecosystems can lead to different targeted effects of the plant protection products in the soil and potential side effects.[9]

Earthworms provide important ecosystem services in agricultural systems through simply burrowing and casting. However, little is known of the extent to which earthworms may have other direct actions that benefit farmers. Work in Austria has suggested that some earthworms may feed on soil-borne plant fungal pathogens and so potentially reduce pathogen infection of crop plants. From both laboratory and field-based experimental work, it has been shown that sclerotia (one stage in the life cycle) of a fungus (*Sclerotinia sclerotiorum* (Lib.) de Bary) that attacks crop plants are consumed directly by *Lumbricus terrestris*, particularly when the sclerotia are fully hydrated. After three months' hydration, more than 60 per cent of sclerotia were consumed in a laboratory experiment. In the field, after eight months the number of recovered sclerotia was lower when earthworms had direct access to them compared with restricted access (fashioned by retaining the fungus in small, earthworm-proof meshes).[10] Further exploration of such feeding relationships could be valuable – the interactions of earthworms, pathogens and cropping systems – particularly as chemical treatments of soils may then be reduced. In addition, earthworm abundance has been demonstrated to increase numbers of above-ground plant pests such as aphids, possibly because earthworms increase the availability of nutrients in the soil, which therefore improves food quality.[11] However, these findings need further investigation in the wider context of complex direct and indirect interactions among soil animal groups and microorganisms.

Investigations of interactions between earthworms and plant pathogens under controlled laboratory conditions.

In an agricultural setting in southern Finland, *L. terrestris* was introduced into an arable production system where it had previously been absent, and its activities monitored over more than a decade.[12] This used an Earthworm Inoculation Unit (EIU) technique, developed for use in hostile soil conditions where all three life stages (adult, cocoon and hatchling) are introduced to enhance long-term colonization success.[13] This work has demonstrated that an ecosystem engineer can be put to good use and in heavy clay soils may act as a natural provider of macropores to assist drainage. Success of introduction and spread was greatest under no-till management and with grassy field margins, where the EIU technique was also employed, acting as a source of additional colonization. Further monitoring, twenty years after inoculation of *L. terrestris*, examined effects on other soil worm groups (other earthworm species, enchytraeids and nematodes) at a localized (midden-scale) and a broader field scale.[14] We found that middens sustained elevated densities of all three faunal groups, but the earthworm community composition was not altered at midden sites. The settlement of *L. terrestris* had no discernible effects on field-scale earthworm and nematode abundances, but enchytraeids were effectively absent beyond the edge of *L. terrestris* colonization into the field, an effect perhaps partially explained by a gradient of increased clay content. These results suggest that *L. terrestris* settlement in a clay soil can significantly increase the spatial patchiness of soil fauna, but may not, except in the case of enchytraeids, affect their field-scale abundances. This is yet another very good example of long-term monitoring, required to reveal important aspects of earthworm-related activity in soils. Further work of this type is also warranted.

Within another realm of agriculture, in the 1980s in southern England, Rothamsted Research investigated earthworms for the breakdown of animal, vegetable and industrial organic wastes through vermicomposting, and was linked with engineering requirements for vermiculture at the National Institute of Agricultural Engineering (NIAE) nearby. In addition, NIAE helped in the development of population models for composting worms, so that the biology, waste usage and hardware associated with the process were all considered in unity along with the economics of vermicompost and protein production. It seemed like the perfect situation, and a related company, British Earthworm Technology (BET), was established to commercialize all aspects of the science. Potentially, it was the ideal set-up and had backing by the UK government, but as mentioned before in a U.S. context, the

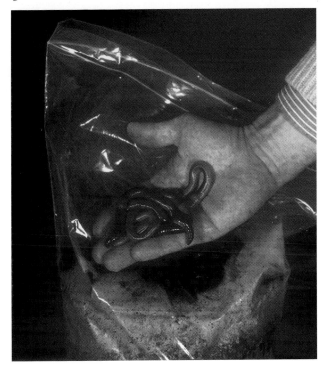

Soil restoration could be enhanced through the addition of earthworms, with care taken to use the most appropriate techniques under potentially hostile soil conditions.

Earthworm Inoculation Units ready for insertion into a semi-restored landfill cap in southern England.

economics did not work for all desired elements and perhaps the engineering was a little too 'robust' for the required biological needs of relatively delicate earthworms.[15] This led to a huge surge in interest in the UK but, with no funding available thereafter, left a void and research in this area in the UK stalled. However, with more recent findings on the details of compost earthworm biology and ecology from groups in other countries, the process of vermicomposting has been shown to have real potential for applied projects, for example in the transformation of wastes from wine production.[16] There is still great scope for employing composting worms in commercial ventures and for encouraging vermicomposting in a circular economy. As a species, we will continue to produce waste organic residues from agriculture, food processing, sewage treatment and some industrial production lines. These materials need to be dealt with carefully and vermicomposting is one solution, but another that can also employ earthworms is within the growing science of soil restoration.

Where land has been despoiled through activities such as open-cast mining or landfill, creating brownfield sites, earthworms can play a part in land restoration. If topsoil has been 'lost' from a site, then earthworms cannot be of use in the first instance, as prior to any biological action there is a need to ensure that the physical and chemical conditions of the soil are at acceptable levels.[17] An example of this can be seen in the restoration of a post-industrial site at a 35-hectare (86 ac) steelworks site in Scotland. The Hallside site near Glasgow was reclaimed from dereliction during the 1990s, when all contaminated material was removed, and a rudimentary soil was formed by spreading stony material from enormous local colliery spoil heaps with locally derived sewage sludge (biosolids) to create a growing medium 2 metres (6½ ft) in depth. This effect-ively sterile substrate was then planted with trees (mainly willow) and *L. terrestris* was introduced, using the EIU technique, plus the addition of turf cut from worm-rich pasture to supply endogeic

Large-scale commercial vermicomposting facilities. Use of such technology may be vital for recycling organic material in the future.

and epigeic species. Earthworm community development has been slow, but after two decades sixteen species are present with respective mean density and biomass of 208 and 71 per square metre.[18] The presence of these earthworms means that they are beginning to provide the required ecosystem services in the soil. The initial crude inorganic and organic mixture, which remains stone-rich, is now a functioning soil, due in large part to the burrowing, ingestion and cast production of the now resident earthworms.

Use of earthworms in soil restoration schemes – harnessing their natural activities to assist reclamation of various habitats degraded by humans – is not new but the practice is growing. The type of degradation, however, may be less severe in some habitats and, for example, can be a function of tree extraction from a forestry site where the resulting problem is soil compaction, caused by the type of machinery used. Laboratory and field-based investigations have shown that the introduction of *L. terrestris* may well be affected by levels of soil compaction, but this earthworm was able to survive at soil bulk densities up to 1.6 grams (1/20 oz) per cubic centimetre. At these higher bulk densities, earthworm burrowing volume was reduced but pores were still created, and tree leaf litter was incorporated into the soil, demonstrating that ecosystem services can be brought back by earthworm addition and help to alleviate the given problems.[19]

Although not a panacea for all soil-related problems, earthworms can assist in the later (biological) stages of soil restoration. Should such work be considered, then a number of questions ought to be asked before any work of this nature is undertaken.[20] Perhaps the most important is: will the addition of earthworms likely result in any gain? Additional questions might be: what are the earthworms expected to do and is it within the scope of their known ecology? If natural colonization will occur from adjacent land, then perhaps inoculation is not required or only additions of deep-burrowing (slow-to-colonize) species may be necessary? Previous research has shown that certain elements are important. The ecological category and combinations of species added (with associated interactions) may be vital and require knowledge of earthworm tolerances with respect to soil physicochemical conditions. The life stages used (cocoon/juvenile/adult) and their origin (collected locally/bred for purpose/purchased) may also be critical to successful establishment and continued survival. Just as quality of inoculum is important, the quantity used is vital but ought to be kept to a minimum while ensuring sustainable population development. Equally there is a need for provision of food (organic matter) for the earthworms, so establishment of a sustainable plant community is vital, often leading to the inclusion of appropriate (nitrogen-fixing, such as *Alnus*) tree planting in combination with earthworm addition.[21] It is essential that the site is monitored, perhaps annually over the first three years and every five years thereafter, or there is no point in this type of soil management. Monitoring of the earthworms themselves – their species, numbers and dispersal – ought to be linked with measurements of developing soil properties.

Modelling of dynamic soil processes is not new and has been used to good effect to consider the fate of organic matter once added to the soil. Soil organic matter (SOM) is key to maintaining soil fertility, mitigating climate change, combatting land degradation, and conserving above- and below-ground biodiversity and associated soil processes and ecosystem services. Microorganisms have always been included in these models, but the role of larger soil organisms has only recently been built into such thinking. Understanding direct and indirect impacts of soil fauna on nutrient availability, carbon sequestration, greenhouse gas emissions and plant growth is key to the understanding of SOM dynamics in the context of global carbon-cycling models. Development of models of SOM are now beginning to more fully incorporate soil organisms and earthworms in particular.[22]

Much has been written here and elsewhere of the positive effects of earthworms in soil processes. But what of greenhouse gas production? From research on vermicomposting in the early 2000s, one published finding that caused great consternation related to nitrogen emissions. It was suggested that gas monitoring showed vermicomposting systems to have the capacity to emit high levels of nitrous oxide (N_2O) and that the earthworms appeared to be primarily responsible. This created an environmental impact of greenhouse gas (GHG) emissions comparable with other agricultural sources and even that of landfill.[23] This research raised issues in the scientific literature and resulted in counterclaims that argued strongly that inadequate control measures were in place and the earlier findings were not a true reflection of vermicomposting output, as comparisons drawn with other GHG-producing industries were scientifically invalid.[24] Nevertheless, as societies are becoming ever more conscious of the environment, further careful monitoring of potential gaseous production may be warranted from such systems, particularly if they become more widely deployed to process organic wastes.

In soil-based systems, earthworms have also been examined for their GHG balance and at times this has been left wanting. Although they are thought to stimulate carbon sequestration in soil aggregates, laboratory experiments showed that earthworms increased carbon dioxide and nitrous oxide emissions, so more recently field trials were set up. N_2O levels were measured in grassland in spring and autumn with two densities of an earthworm species (*Lumbricus rubellus*) with additions of fertilizers. Results from autumn showed significant increases in GHG emissions at both earthworm densities. The results suggest that the pathways through which earthworms affect nitrogen cycling (and thereby N_2O emission) differ with weather conditions – dry spring weather overruled any earthworm effects, whereas in autumn earthworms mainly improved soil aeration and thereby increased both plant nitrogen uptake and the diffusion of N_2O to the atmosphere. Earthworms do not emit N_2O themselves, but rather affect the microbial processes that produce and consume this gas in the soil through their activity. The intensity of the earthworm effect in the field is a function of soil physicochemical parameters and thereby influenced by meteorological and seasonal dynamics.[25] Further investigations are required to obtain a wider view of the earthworm-related influence in the sphere of GHG emissions with a more global perspective.

Just as at local levels – for example, UK earthworm distribution – little is known about global earthworm distribution, diversity and even threats to existence. A recent global dataset of sampled earthworm communities, however, from some 7,000 sites in 57 countries was used as a basis for predicting patterns in earthworm diversity, abundance and biomass.[26] Species richness and abundance was found to typically peak at higher latitudes, patterns that were not expected and the opposite of those observed in most above-ground organisms. Surprisingly, climate variables such as precipitation were found to be more important at shaping earthworm communities than soil properties or habitat. These findings suggest that the challenges that we face through climate change may substantially alter earthworm communities with cascading effects on other species in these ecosystems. There is a need to be aware of the bigger picture in future biodiversity conservation and broaden consideration beyond above-ground habitats and flagship species. Smaller organisms also matter and their often less obvious contributions to ecosystem function may even be more important than larger animals around them.

Adult *Dendrobaena pygmaea,* a very small British worm, often overlooked due to size but worthy of much more investigation.

Two forms (pink and green) of the earthworm *Allolobophora chlorotica*, which have been investigated for relatedness through DNA analysis and mating experiments.

We have already touched upon the use of molecular techniques for scientific investigation with respect to the identity of the 'Croston Worm', nevertheless this type of analysis is just as relevant and current with earthworms. Over the past twenty years, DNA 'barcoding' has become more reliable, less expensive and simpler to use. Molecular data are therefore now considered as complementary diagnostic characters that can be used alongside traditional earthworm species identification.[27] It is now possible to differentiate between species plasticity, which is the ability to adapt to given environmental conditions, and real (sub)species differences. For example, a DNA investigation of the green earthworm (*Allolobophora chlorotica*) showed five highly divergent lineages, suggesting the presence of multiple cryptic species.[28] This information is of importance as it means that comparisons drawn between historical studies now need to consider whether the same lineage was under investigation, as each may exhibit different biological traits. Even *Lumbricus terrestris* has been given closer scrutiny of late and it has been determined that this species ought to be reconsidered, as significant differences in DNA show two lineages, equivalent to those found between two very different earthworm species. Here, however, the two forms of *L. terrestris* are morphologically indistinguishable, except by overlapping size-related characteristics, and the smaller species is now named *Lumbricus herculeus*.[29] Once again, this calls into question previously reported research on nominal *L. terrestris*, which could have been conducted on either or both species, although *L. herculeus* has yet to be found outside Europe.

With forensic-like investigation, collection of DNA samples from a given species allows for direct comparisons to be drawn from where it occurs naturally and from where it has been introduced, often to new continents. In this way it has been possible, for example, to show that the European earthworms present in North America were derived from multiple introductions and a variety of origins.[30] Such biological invasions can have positive effects in some agro-ecosystems, but are now more frequently shown to be deleterious to the ecosystem in which they occur. There are currently no effective methods to eradicate established earthworm populations without unacceptable non-target effects, so the main technique for managing invasions is further prevention through education of likely dispersal agencies.[31]

DNA analysis of earthworms has yet another strand for exploration. From environmental DNA (eDNA) analysis, it is possible to discover which species (animal, plant or microbe) are present in a given habitat, provided that reference material is available for comparison. A recent study has suggested that we may no longer need to collect earthworms in more traditional ways, as previously described, but may simply be able to take soil samples and extract the traces of DNA left by earthworms to establish which species are present and provide a level of species richness in a given area.[32] Results have been compared with more traditional sampling methods and give a good match, although number and biomass per square metre cannot be gained from this technique. Nevertheless, eDNA analysis may provide the possibility of rapid assessment of earthworm communities from multiple localities, enabling stimulation of further ecological research on these organisms, as appropriate.

So just as we are bringing about changes to the climate on Earth, through habitat destruction, deforestation, pollution, fossil

Earthworm cocoons in Petri dishes ready for experiments with freezing to investigate the potential of an earthworm 'biobank'.

fuel consumption and GHG emissions, to name just a few activities, we are also directly affecting distribution patterns of earthworms and, as already mentioned, affecting temperate forests in North America. In addition, research in the boreal ecosystems of Canada has demonstrated a rise in soil temperature that has allowed for a northern migration of earthworm species. This has enabled novel soil bioturbation by the action of ecosystem engineering earthworms and exploitation of organic matter, once locked up in permafrost.[33] This is causing greater GHG emissions through the interaction of earthworms and microorganisms in the soil, which in turn will lead to further climate change. Through DNA analysis we may be able to establish where earthworms have originated from, after inadvertently assisting their global redistribution, but is this of any significance now that some have joined a cycle of events that may be unstoppable?

Globally, scientists, governments and industry must begin to seriously address the issues of climate change that affect ourselves, earthworms and all living organisms on Earth. One set of thinking in the biological sphere is that the creation of a reservoir of biodiversity in some form is required. This is already underway in many areas, such as a seedbank for plants, as held at Kew Gardens in London,[34] or perhaps part of a frozen ark, as under consideration and development at the Natural History Museum in London.[35] As a part of the latter, an 'Earthworm Bank' has been discussed and experiments with freezing and viability testing of cocoons have begun. In the long term this may act as a living source for stock populations of common and rare species for research and educational work or even repopulation, as required in the future, using trusted inoculation techniques.

Worms have formed a part of folklore for millennia and will continue to do so. We have learned much of their natural history over the past few centuries, but, as suggested throughout this book, there is still so much to be discovered. Our historical and current environmental activities may have opened a can of worms regarding Earth's climate future and we are now at a delicate point in our history. Perhaps now, to help us move forwards, we can further harness the ecosystem services offered by earthworms, possibly the most important but underappreciated group of organisms on our planet.

Appendix: Phrases and Words Referring to Worms

BOOKWORM A person who enjoys reading (but also a beetle larva that eats paper)

EARLY BIRD CATCHES THE WORM If you want to succeed, start as soon as possible

EARWORM (BRAIN WORM) A catchy piece of music that you may hear and then cannot get out of your head – it keeps playing over and over – and can become annoying

EVEN A WORM WILL TURN Even submissive people will be angered and retaliate if they are continually agitated or oppressed. For example, as used by William Shakespeare (*Henry VI, Part 3*, II.2.15–18):

Who 'scapes the lurking serpent's mortal sting?
Not he that sets his foot upon her back.
The smallest worm will turn being trodden on,
And doves will peck in safeguard of their brood.

This was probably adopted from the English writer John Heywood, who included 'Treade a worme on the tayle, and it must turne agayne' in his *A Dialogue conteinyng the nomber in effect of all the Prouerbes in the Englishe tongue* (London, 1546)

GUMMY WORMS Sweets, often gelatine-based, simply shaped to look like earthworms. Fun can be had by making children think they are eating real worms (parents often put them into party foods). Can be bought or made with many recipes available

HAVE YOU GOT WORMS? Often a reference to people who seem to eat more than might be expected (feeding a tapeworm?). It can be quite funny (or so I think) should my wife say this to me, as I reply, 'Yes, I have a whole laboratory full!'

OPEN A CAN OF WORMS To (inadvertently) create numerous new problems while trying to solve one. Its origins are uncertain but may stem from the Greek myth of the opening of Pandora's box. This myth held that, even though warned not to do so, the curious Pandora opened the box or jar and released all the world's evils, leaving only hope inside

THE WORM Formerly a popular name for an array of ailments supposedly caused by the work of a worm in various parts of the human body

WORM A contemptible person

WORM Write Once, Read Many (computing)

WORM'S-EYE VIEW Viewpoint of a worm. Can be thought of as the opposite of a bird's -eye view

WORM JOKE What's worse than biting into an apple and finding a worm? Finding half a worm!

WORM-MOON The full moon that appears in March. Indigenous Americans called this last full moon of winter the Worm Moon after the worm trails that would appear in the newly thawed ground

WORM ONE'S WAY INTO SOME PLACE Obtain a position or situation in a sly or cunning manner

WORMSTONE Part of an experiment set up at Down House by
Charles Darwin to measure the action of earthworms in the
soil

WORMWOOD (*ARTEMISIA ABSINTHIUM*) A bitter herb, native
to Britain but introduced to America, known as an ingredient
in the drink absinthe. It was believed to be useful for
expelling parasitic worms from the intestine, hence the
name. In moderation, wormwood may still offer numerous
benefits include pain relief and reduced inflammation.
The nurse in Shakespeare's *Romeo and Juliet* refers to its use,
'for I had then laid wormwood to my dug' (1.3.26), meaning
that she had weaned Juliet, then aged three, by using the
bitter taste of wormwood on her nipple. (Not to be confused
with 'wormwood to my dog', a way of seeking to rid a canid
of intestinal worms using *A. absinthium*, as already suggested.)

Timeline of the Worm

555 MILLION BP	200–140 MILLION BP	*c.* AD 60

Marine worms as small as a grain of rice living in sediments

Earthworms colonize the land

Pliny the Elder describes large parasitic worms breeding in the human gut

1789	1800S	1881

Gilbert White first describes the ecosystem services provided by earthworms in *The Natural History of Selborne*

Earthworms regarded as garden pests along with slugs and snails

Charles Darwin publishes a seminal earthworm book, *The Formation of Vegetable Mould Through the Action of Worms*

1980S		1988

Worm charming becomes a popular sport in England

Nematode worm *Caenorhabditis elegans* is the first multicellular organism to have its complete genome sequenced

1600S	1668	1758

In Europe, earthworm oil is regarded as a pain reliever

Francesco Redi discovers the life cycle of tapeworms infecting humans and intermediary hosts

Linnaeus names *Lumbricus terrestris*, *Ascaris lumbricoides* and four other worm species, which he groups as 'Vermes'

1896	1940S	1970S

The Cambridge Natural History, vol. II, explores the diversity of worm varieties

Earthworm ecology studied in detail at Rothamsted Experimental Station

Vermicomposting of organic waste becomes popular in the USA and parts of Europe

	2018	2022

Nematode worms sent to visit the International Space Station

12th International Symposium on Earthworm Ecology (ISEE12) held in Rennes, France

References

1 INTRODUCING THE WORM

1 Kevin R. Butt and N. Grigoropoulou, 'Basic Research
  Tools for Earthworm Ecology', *Applied and Environmental
  Soil Science*, 29 November 2009, www.hindawi.com.

2 Kevin R. Butt et al., 'Field and Laboratory Investigations
  of *Lumbricus badensis* Ecology and Behaviour', *Pedosphere*,
  XXI/3 (2021), pp. 471–4.

3 Carolus Linnaeus [Carl Von Linné], *Systema Naturae*,
  10th edn (Uppsala, 1758).

4 Daniel Le Clerc, *A Natural and Medicinal History of Worms,
  Bred in the Bodies of Men and Other Animals* (London, 1721).

5 Francesco Redi, *Esperienze intorno alla generazione degl'insetti*
  (Florence, 1668).

6 F.E.G. Cox, 'History of Human Parasitology', *Clinical Microbial
  Reviews*, XV/4 (2002), pp. 595–612.

7 Robert Hooke, *Micrographia; or, Some Physiological Descriptions
  of Minute Bodies Made by Magnifying Glasses* (London, 1665),
  pp. 208–10.

8 Antony van Leeuwenhoek, 'Part of a Letter from
  Mr Anthony van Leeuwenhoek, Concerning the Worms
  in Sheeps Livers, Gnats and Animalcula in the Excrements
  of Frogs', *Philosophical Transactions of the Royal Society of London*,
  XXII (1700), pp. 509–18.

9 Lord Byron, 'The Bride of Abydos' (1813), available at
  www.best-poems.net, accessed 21 August 2022.

10 William Blake, 'The Sick Rose', *Songs of Experience* (London, 1794).

11 Anon, 'A Disputation Between the Body and Worms', 15th-century poem, British Library, Add. Ms 37049, 32*v*.

12 Edgar Allan Poe, 'The Conqueror Worm', *Graham's Lady's and Gentleman's Magazine*, ccv (January 1843).

13 S. F. Harmer and A. E. Shipley, eds, *The Cambridge Natural History*, vol. ii (London, 1896).

14 Jan Kozlowski, M. Jaskulska and M. Kozlowska, 'Evaluation of the Effectiveness of Iron Phosphate and the Parasitic Nematode *Phasmarhabditis hermaphrodita* in Reducing Plant Damage Caused by the Slug *Arion vulgaris Moquin-Tandon*, 1885', *Folia Malacologica*, xxii/4 (2014), pp. 293–300.

15 *Caenorhabditis elegans*, www.wormbase.org, accessed 4 September 2021.

16 'Worms in Space: The Molecular Muscle Experiment', uk Space Agency, www.gov.uk, accessed 3 November 2022.

17 'Parasites – Ascariasis', Centers for Disease Control and Prevention, www.cdc.gov, accessed 4 September 2021.

18 'Soil-Transmitted Helminthiases', World Health Organization, www.who.int, accessed 4 September 2021.

19 Steve Ott, 'Worms You Don't Want', *Kitchen Garden*, 30 August 2018, www.kitchengarden.co.uk.

20 'Hand Gathering Fisheries', Southern Inshore Fisheries and Conservation Authority, www.southern-ifca.gov.uk, accessed 4 September 2021.

21 Andres R. Acosta-Galvis, Mauricio Torres and Paola Pulido-Santacruz, 'A New Species of *Caecilia* (Gymnophiona: Caeciliidae) from the Magdalena Valley Region of Colombia', *Zookeys*, 884 (2019), pp. 135–57.

22 Silk Thread Containing Spider Thread Protein and Silkworm Producing the Silk Thread. United States Patent Application 20080287651, filed 12 January 2005, www.freepatentsonline.co.uk, accessed 4 September 2021.

23 Julián Monge-Nájera, 'Onychophorology, the Study of Velvet Worms, Historical Trends, Landmarks, and Researchers from 1826 to 2020 (A Literature Review)', *Uniciencia*, xxxv/1 (2021), pp. 210–30.

24 Beverley Van Praagh, 'The Biology and Conservation of the Giant Gippsland Earthworm *Megascolides australis* Mccoy, 1878', *Soil Biology and Biochemistry*, XXIV/12 (1992), pp. 1363–7.

25 Siobhan M. Quigg et al., 'A Re-Examination of the Taxonomic Status of *Prostoma jenningsi* – a Freshwater Nemertean', *Zootaxa*, MMMMDCCXXII/2 (2020), pp. 175–84.

26 Charles Darwin, *The Formation of Vegetable Mould through the Action of Worms with Observations on Their Habits* (London, 1881), p. 60.

27 Samuel W. James, 'Revision of the Earthworm Genus *Archipheretima* Michaelsen (Clitellata: Megascolecidae), with Descriptions of New Species from Luzon and Catanduanes Islands, Philippines', *Organisms, Diversity and Evolution*, IX/3 (2009), pp. 244.e1–244.e16.

28 Hong Chen et al., 'A Cambrian Crown Annelid Reconciles Phylogenomics and the Fossil Record', *Nature*, 583 (2020), pp. 249–52.

29 Scott D. Evans et al., 'Discovery of the Oldest Bilaterian from the Ediacaran of South Australia', *PNAS*, CXVII/14 (2020), pp. 7845–50.

30 Trevor G. Piearce, K. Oates and W. J. Carruthers, 'A Fossil Earthworm Embryo (Oligochaeta) from Beneath a Late Bronze Age Midden at Potterne, Wiltshire', *Journal of Zoology*, CCXX/4 (1990), pp. 537–42.

31 Karen Chin, D. Pearson and A. A. Ekdale, 'Fossil Worm Burrows Reveal Very Early Terrestrial Animal Activity and Shed Light on Trophic Resources After the End-Cretaceous Mass Extinction', *PLOS ONE*, VIII/8 (2013), p. E70920.

32 Mariano Verde et al., 'A New Earthworm Fossil from Paleosols: Aestivation Chambers from the Late Pleistocene Sopas Formation of Uruguay', *Palaeogeography, Palaeoclimatology, Palaeoecology*, CCXLIII/3–4 (2007), pp. 339–47.

33 Youri Lammers et al., 'Clitellate Worms (Annelida) in Late Glacial and Holocene Sedimentary DNA Records from the Polar Urals and Northern Norway', *Boreas*, XLVIII/2 (2018), pp. 317–29.

34 Aristotle (attrib.), 'Aristotle Quotes on Earth', *Today in Science History*, https://todayinsci.com, accessed 4 September 2021.

1 H. Webb and M. A. Grigg, *Modern Science* (Cambridge, 1949), Book II, pp. 119–23.
2 Emma Sherlock, *Key to the Earthworms of Britain and Ireland*, 2nd edn (Telford, 2018).
3 Clive A. Edwards and N. Q. Aracon, 'The Use of Earthworms in the Breakdown of Organic Wastes to Produce Vermicomposts and Animal Feed Protein', in *Earthworm Ecology*, ed. Clive A. Edwards (Boca Raton, FL, 2004), pp. 345–79.
4 Ronald E. Gaddie Sr and D. E. Douglas, *Earthworms for Ecology and Profit*, vol. II: *Earthworms and the Ecology* (Ontario, CA, 1977).
5 Mary Appelhof, *Worms Eat My Garbage* (Kalamazoo, MI, 1997).
6 Kevin R. Butt and B. Williams, 'Vermiculture and Vermicomposting in the United Kingdom', in *Vermiculture Technology*, ed. C. A. Edwards, N. Q. Arancon and R. Sherman (Boca Raton, FL, 2011), pp. 423–36.
7 'Vermicomposting Toilets', www.vermicompostingtoilets.net, accessed 4 September 2021.
8 Clare Furlong et al., 'Is Composting Worm Availability the Main Barrier to Large-Scale Adoption of Worm-Based Organic Waste Processing Technologies?', *Journal of Cleaner Production*, CLXIV (2017), pp. 1026–33.
9 H. H. Thomas, ed., *Gardening for Amateurs: A Simple Complete, and Practical Guide for Garden Lovers*, vol. I (New York, 1915).
10 'Talpirid Mole Bait', Bell Laboratories, Inc., www.belllabs.com, accessed 4 September 2021.
11 'Blackawton International Festival of Wormcharming', Blackawton Community Website, 18 February 2020, https://blackawtoncommunity.com.
12 K.E.L. Simmons, 'Foot-Movements in Plovers and Other Birds', *British Birds*, LIV (1961), pp. 34–9.
13 J. H. Kaufmann, 'Stomping for Earthworms by Wood Turtles, *Clemmys insculpta*: A Newly Discovered Foraging Technique', *Copeia*, 4 (1986), pp. 1001–4.
14 Daniel Carpenter et al., 'Mapping of Earthworm Distribution for the British Isles and Eire Highlights the Under-Recording of an

Ecologically Important Group', *Biodiversity and Conservation*, 21 (2011), pp. 475–85.

15 'Soil and Earthworm Survey', OPAL: Citizen Science for Everyone, Imperial College London, www.imperial.ac.uk, accessed 4 September 2021.

16 The Earthworm Society of Britain, www.earthwormsoc.org.uk, accessed 4 September 2021.

17 Kevin R. Butt and N. Grigoropoulou, 'Basic Research Tools for Earthworm Ecology', *Applied and Environmental Soil Science*, 29 November 2009, www.hindawi.com.

18 Reginald W. Sims and B. M. Gerard, *Earthworms* (Shrewsbury, 1999).

19 U. Thielemann, 'Elektrischer Regenwurmfang Mit Der Oktette-Methode', *Pedobiologia*, XXIX (1986), pp. 296–302.

20 Alfred R. Wallace, 'On the Insects Used as Food in the Indians of the Amazon', *Royal Entomological Society London*, II (1853), pp. 241–4.

21 Maurizio G. Paoletti et al., 'Nutrient Content of Earthworms Consumed by Ye'Kuana Amerindians of the Alto Orinoco of Venezuela', *Proceedings of the Royal Society London B*, CCLXX (2003), pp. 249–57.

22 Gaddie and Douglas, *Earthworms for Ecology and Profit*, vol. II.

23 Competition and Markets Authority. 'Pyramid Selling: Advice for the Public and Communities', 13 October 2014, www.gov.uk, accessed 4 September 2021.

24 Zhenjun Sun, 'Earthworm as a Biopharmaceutical: From Traditional to Precise', *European Journal of Biomedical Science*, I/2 (2015), pp. 28–35.

25 Alok Atreya and S. Aryal, 'Does Eating Earthworms Increase Breast Milk?', *Journal of Gynecology Obstetrics and Human Reproduction*, I/3 (2021), no. 102170.

26 Marta J. Fiołka et al., 'Antitumor Activity and Apoptotic Action of CoelomicFluid from the Earthworm *Dendrobaena veneta* Against A549 Human Lung Cancer Cells', *Apmis*, CXXVII/6 (2019), pp. 435–48.

1 Nico M. van Straalen, 'Evolutionary Terrestrialization Scenarios for Soil Invertebrates', *Pedobiologia*, LXXXVII–LXXXVIII (2021), no. 150753.

2 John C. Loudon, *Encyclopaedia of Gardening* (London, 1826), p. 436.

3 Ibid., p. 437.

4 Ibid.

5 Gilbert White, *The Natural History of Selborne* (London, 1789), Letter XXXV.

6 Ibid.

7 Ibid.

8 Charles Darwin, *The Formation of Vegetable Mould through the Action of Worms with Observations on Their Habits* (London, 1881).

9 Christian Feller et al., 'Charles Darwin, Earthworms and the Natural Sciences: Various Lessons from Past to Future', *Agriculture, Ecosystem and Environment*, XCIX/1–3 (2003), pp. 29–49.

10 Charles Darwin, *On the Origin of Species* (London, 1859).

11 Darwin, *The Formation of Vegetable Mould*, p. 288.

12 Ibid., p. 7.

13 Kevin R. Butt et al., 'Darwin's Earthworms Revisited', *European Journal of Soil Biology*, XLIV/3 (2008), pp. 255–9.

14 Darwin, *The Formation of Vegetable Mould*, p. 149.

15 Ibid., p. 148.

16 Kevin R. Butt, J. Frederickson and R. M. Morris, 'Investigations of an Earthworm Inoculation Experiment, London Borough of Hillingdon', *Waste Planning*, 7 (1993), pp. 9–12.

17 Clive A. Edwards and P. J. Bohlen, *Biology and Ecology of Earthworms* (London, 1996), pp. 97–9.

18 Butt et al., 'Darwin's Earthworms Revisited', pp. 255–9.

19 Clive G. Jones, J. H. Lawton and M. Shachak, 'Organisms as Ecosystem Engineers', *Oikos*, LXIX/3 (1994), pp. 373–86.

20 Martin J. Shipitalo and R.-C. Le Bayon, 'Quantifying the Effects of Earthworms on Soil Aggregation and Porosity', in *Earthworm Ecology*, ed. Clive A. Edwards (Boca Raton, FL 2004), pp. 183–200.

21 Ken E. Lee, *Earthworms: Their Ecology and Relationships with Soils and Land Use* (Sydney, 1985).

22 James P. Curry and O. Schmidt, 'The Feeding Ecology of Earthworms – A Review', *Pedobiologia*, L/6 (2007), pp. 463–77.

23 Ibid.

24 F. Raw, 'Studies of Earthworm Populations in Orchards, I: Leaf Burial in Apple Orchards', *Annals of Applied Biology*, L/3 (1962), pp. 389–404.

25 Chris N. Lowe and K. R. Butt, 'Growth of Hatchling Earthworms in the Presence of Adults: Interactions in Laboratory Culture', *Biology and Fertility of Soils*, 35 (2002), pp. 204–9.

26 Alexei V. Uvarov, 'Inter- and Intraspecific Interactions in Lumbricid Earthworms: Their Role for Earthworm Performance and Ecosystem Functioning', *Pedobiologia*, LIII/1 (2009), pp. 1–27.

27 Chris N. Lowe and K. R. Butt, 'Earthworm Culture, Maintenance and Species Selection in Chronic Ecotoxicological Studies: A Critical Review', *European Journal of Soil Biology*, XLIII (2007), pp. 281–8.

28 Darwin, *The Formation of Vegetable Mould*, p. 62.

29 Ibid., p. 78.

30 This is something that Jimmy Doherty and I re-enacted at Down House for a BBC TV programme in 2009, and the results we obtained using paper triangles covered with lard were similar to those reported by Darwin. *Jimmy Doherty in Darwin's Garden*, episode 3, 'Of Apes and Men', BBC TV, 2009, limited availability at www.open.ac.uk, accessed 4 September 2021.

31 John E. Satchell and D. G. Lowe, 'Selection of Leaf Litter by *Lumbricus terrestris*', in *Progress in Soil Biology*, ed. Otto Graff and J. E. Satchell (Amsterdam, 1967), pp. 102–19.

32 Nalika S. S. Rajapaksha et al., 'Earthworm Selection of Short Rotation Forestry Leaf Litter Assessed through Preference Testing and Direct Observation', *Soil Biology and Biochemistry*, LXVII (2013), pp. 12–19.

33 Sandra A. Moody et al., 'Selective Consumption of Decomposing Wheat Straw by Earthworms', *Soil Biology and Biochemistry*, XXVII/9 (1995), pp. 1209–13.

34 Doyen T. T. Hoang et al., 'Hotspots of Microbial Activity Induced by Earthworm Burrows, Old Root Channels, and Their Combination', *Biology and Fertility of Soils*, 52 (2016), pp. 1105–19.

35 Dolores Trigo et al., 'Mutualism Between Earthworms and Microflora', *Pedobiologia*, 43 (1999), pp. 866–73.

36 John E. Satchell, 'Earthworm Microbiology', in *Earthworm Ecology from Darwin to Vermiculture*, ed. J. E. Satchell (London, 1983), pp. 351–64.

37 Lucio Montecchio et al., 'Potential Spread of Forest Soil-Borne Fungi through Earthworm Consumption and Casting', *iforest-Biogeosciences and Forestry*, VIII/3 (2015), pp. 295–301.

38 Kevin Hoeffner et al.,'Two Distinct Ecological Behaviours within Anecic Earthworm Species in Temperate Climates', *European Journal of Soil Biology*, CXIII (2022), no. 103446.

39 Darwin, *The Formation of Vegetable Mould*, p. 283.

40 Philip Greenwood, 'A Prototype Tracing-Technique to Assess the Mobility of Dispersed Earthworm Casts on a Cultivated Hillslope Soil Under Unconsolidated and Crusted Surface Conditions', *Geoderma*, CD (2021), no. 115220.

41 Darwin, *The Formation of Vegetable Mould*, p. 133.

42 Kevin R. Butt et al., 'Action of Earthworms on Flint Burial – A Return to Darwin's Estate', *Applied Soil Ecology*, CIV (2016), pp. 157–62.

43 Arthur Keith, 'A Postscript to Darwin's Formation of Vegetable Mould through the Action of Worms', *Nature*, CLXIX (1942), pp. 716–20.

44 André Kretzschmar, 'Description des galeries de vers de terre et variation saisonnière des réseaux (observations en conditions naturelles)', *Revue d'ecologie et de biologie du sol*, XIX/4 (1982), pp. 579–91.

45 Martin J. Shipitalo and K. R. Butt, 'Occupancy and Geometrical Properties of *Lumbricus terrestris* L. Burrows Affecting Infiltration', *Pedobiologia*, 43 (1999), pp. 782–94.

46 Reginald W. Sims and B. M. Gerard, *Earthworms* (Shrewsbury, 1999), p. 92.

47 Grey T. Coupland and J. I. Mcdonald, 'Extraordinarily High Earthworm Abundance in Deposits of Marine Macrodetritus Along Two Semi-Arid Beaches', *Marine Ecology Progress Series*, 361 (2008), pp. 181–9.

48 Manuel Blouin et al., 'A Review of Earthworm Impact on Soil Function and Ecosystem Services', *European Journal of Soil Science*, LXIV/2 (2013), pp. 161–82.
49 Darwin, *The Formation of Vegetable Mould*, p. 287.
50 Ibid., p. 13.
51 Gerhard Cadee, 'Gilbert White and Darwin's Worms', *Ichnos*, X/1 (2003), pp. 47–9.

4 ASIDE FROM SCIENCE

1 Charles Darwin, *The Formation of Vegetable Mould through the Action of Worms with Observations on Their Habits* (London, 1881), p. 34.
2 Kevin R. Butt, C. N. Lowe and P. Duncanson, 'Earthworms of an Urban Cemetery in Preston: General Survey and Burrowing of *Lumbricus terrestris*', *Zeszyty Naukowe*, 17 (2014), pp. 23–30.
3 A. Sutherland, '"The Battle of the Tooth Worm": Strange Ivory Carving', *Ancient Pages*, 11 August 2015, www.ancientpages.com, accessed 4 September 2021.
4 University of Maryland, Baltimore, 'Do You Believe in "Tooth Worms?" Micro-Images of Strange, Worm-Like Structures Uncovered Inside Dissected Molar', *ScienceDaily*, 28 July 2009, www.sciencedaily.com, accessed 4 September 2021.
5 'Prevented Toothache', *Richmond* [KY] *Climax*, 24 May 1911.
6 'Syrup Jar for Oil of Earthworms, Italy, 1731–1770', Science Museum Group, https://collection.sciencemuseumgroup.org.uk, accessed 4 September 2021.
7 John Quincy, *Pharmacopoeia Officinalis and Extemporanea* (London, 1719), p. 448.
8 Dave Stogie, 'Lambton Worm Song', www.youtube.com, accessed 4 September 2021.
9 Roy Chapman Andrews, *On the Trail of Ancient Man* (New York, 1926), p. 103.
10 'Monsters of the Mind: Mongolian Death Worm', *Weird n' Wild Creatures Wiki*, https://weirdnwildcreatures.fandom.com, accessed 4 November 2022.

11 Gary Larson, *There's a Hair in My Dirt! A Worm's Story* (New York, 1998).

12 Roger Hargreaves and Adam Hargreaves, *Walter the Worm* (London, 2018).

13 Roald Dahl, *The Twits* (London, 1980).

14 'Live Stunt Opens Can of Worms', *The Guardian*, 2 November 1991.

15 Roald Dahl, *James and the Giant Peach* (New York, 1961).

16 *Earthworm Jim* (1995–6), www.imdb.com, accessed 4 September 2021.

17 'Chap.4. The Earthworm's Monologue', www.youtube.com, accessed 4 September 2021.

18 Anne Sexton, 'Earthworm', *45 Mercy Street* (Boston, MA, 1976)

19 Donna Word Chappell, 'The Earthworm', available at www.ellenbailey.com, accessed 4 September 2021.

20 Ted Hughes, *Crow* (London, 1970), p. 10.

21 John Gay, 'The Ravens, the Sexton and the Earth-Worm', *Fables*, vol. II (London, 1738).

22 William Butler Yeats, 'The Man Who Dreamed of Faeryland', *The Rose* (London, 1893).

23 Kevin R. Butt, 'The Effects of Temperature on the Intensive Production of *Lumbricus terrestris* L. (Oligochaeta: Lumbricidae)', *Pedobiologia*, XXXV (1991), pp. 257–64.

24 Maciej Kocinski et al., 'Experimental Induction of Autotomy in Two Potential Model Lumbricid Earthworms *Eisenia andrei* and *Aporrectodea caliginosa*', *Invertebrate Survival Journal*, XXXIII/1 (2016), pp. 11–17.

25 Andrew R. Gehrke et al., 'Acoel Genome Reveals the Regulatory Landscape of Whole-Body Regeneration', *Science*, CCCLXIII/6432 (2019), pp. eaau6173.

26 Raffaele Gaeta, F. Bruschi and V. Giuffra, 'The Painting of St. Roch in the Picture Gallery of Bari (15th Century): An Ancient Representation of Dracunculiasis?', *Journal of Infection*, LXXIV/5 (2017), pp. 519–21.

27 'Dracunculiasis (Guinea-Worm Disease)', World Health Organization, www.who.int, accessed 4 September 2021.

28 'Parasites – Guinea Worm', Centers for Disease Control and Prevention, www.cdc.gov, accessed 4 September 2021.

29 G. L. Wood, ed., *The Guinness Book of Animal Facts and Feats*, 2nd edn (London, 1976).

30 Myka Baum, www.mykabaum.com, accessed 4 September 2021.

31 A. L. Brown, *Ecology of Soil Organisms* (Portsmouth, NH, 1978), p. 67.

32 Darwin, *The Formation of Vegetable Mould*, p. 55.

33 Kundong D. Wang and G. Z. Yan, 'An Earthworm-Like Microrobot for Colonoscopy', *Biomedical Instrumentation and Technology*, 40 (2006), pp. 471–8.

34 Akira Oyama et al., 'Detection of Rust from Images in Pipes Using Deep Learning', *18th International Conference on Ubiquitous Robots (UR)*, 12–14 July 2021, pp. 476–9, available at https://ieeexplore. ieee.org.

35 'Ge Research to Demonstrate Giant Earthworm-Like Robot for Superfast, Ultra-Efficient Tunnel Digging', Ge Research, 20 May 2020, www.ge.com, accessed 4 November 2022.

36 Deepak Trivedi et al., 'Soft Robotics: Biological Inspiration, State of the Art and Future Research', *Applied Bionics and Biomechanics*, V/3 (2008), pp. 99–117.

37 'Worms Join Man United', *Sunday People*, 8 November 1998.

5 *LUMBRICUS TERRESTRIS*: (NOT SUCH) A COMMON EARTHWORM

1 Carolus Linnaeus [Carl Von Linné], *Systema Naturae*, 10th edn (Uppsala, 1758).

2 Reginald W. Sims and B. M. Gerard, *Earthworms* (Shrewsbury, 1999), p. 108.

3 Izaak Walton, *The Compleat Angler* (London, 1653).

4 Visa Nuutinen and K. R. Butt, 'Homing Behaviour of the Earthworm *Lumbricus terrestris* L. in the Laboratory', *Soil Biology and Biochemistry*, 37 (2005), pp. 805–7.

5 Hans Kruuk, 'Foraging and Spatial Organisation of the European Badger *Meles meles* L.', *Behavioural Ecology and Sociobiology*, IV/1 (1978), pp. 75–89.

6 Kevin R. Butt and V. Nuutinen, 'The Dawn of the Dew Worm', *Biologist*, LII/4 (2005), pp. 218–23.

7 *Lumbricus terrestris (Lumbricidae) – Locomotion and Feeding Habit* (1973), www.filmarchives-online.eu, accessed 4 September 2021.

8 Richard Dawkins, *The Extended Phenotype: The Long Reach of the Gene* (Oxford, 1982).

9 Charles Darwin, *The Formation of Vegetable Mould through the Action of Worms with Observations on Their Habits* (London, 1881), p. 59.

10 R. Andrew King et al., 'Prey Choice by Carabid Beetles Feeding on an Earthworm Community Analysed Using Species- and Lineage-Specific Pcr Primers', *Molecular Ecology*, XIX/8 (2010), pp. 1721–32.

11 Darwin, *The Formation of Vegetable Mould*, p. 60.

12 Martin J. Shipitalo, V. Nuutinen and K. R. Butt, 'Interaction of Earthworm Burrows and Cracks in a Clayey, Subsurface-Drained Soil', *Applied Soil Ecology*, XXVI (2004), pp. 209–17.

13 Reginald W. Sims and B. M. Gerard, *Earthworms* (Shrewsbury, 1999), p. 28.

14 Visa Nuutinen et al., 'Abundance of the Earthworm *Lumbricus terrestris* L. in Relation to Subsurface Drainage Pattern on a Sandy Clay Field', *European Journal of Soil Biology,* XXXVII (2001), pp. 310–14.

15 Visa Nuutinen and Kevin R. Butt, 'Interaction of *Lumbricus terrestris* L. Burrows with Field Subdrains', *Pedobiologia*, XLVII/5–6 (2003), pp. 578–81.

16 Mervi Nieminen et al., 'Local Land Use Effects and Regional Environmental Limits on Earthworm Communities in Finnish Arable Landscapes', *Ecological Applications*, XXI/8 (2011), pp. 3162–77.

17 Martin Holmstrup and Karl E. Zachariassen, 'Physiology of Cold Hardiness in Earthworms', *Comparative Physiology and Biochemistry Part A: Physiology*, CXV/2 (1996), pp. 91–101.

18 Visa Nuutinen and K. R. Butt, 'Worms from the Cold: Lumbricid Life Stages in Boreal Clay During Frost', *Soil Biology and Biochemistry*, XLVII/7 (2009), pp. 1580–82.

19 Visa Nuutinen et al., 'Dew-Worms in White Nights: High-Latitude Light Constrains Earthworm (*Lumbricus terrestris*) Behaviour at the Soil Surface', *Soil Biology and Biochemistry,* LXXII (2014), pp. 66–74.

20 Kevin R. Butt, V. Nuutinen and T. Sirén, 'Resource Distribution and Surface Activity of Adult *Lumbricus terrestris* L. in an Experimental System', *Pedobiologia*, 47 (2003), pp. 548–53.

21 Michael S. Laverack, *The Physiology of Earthworms* (Oxford, 1963).

22 John A. Wallwork, *Earthworm Biology* (London, 1983).

23 Darwin, *The Formation of Vegetable Mould*, p. 24.

24 Visa Nuutinen and K. R. Butt, 'Pre-Mating Behaviour of the Earthworm *Lumbricus terrestris* L.', *Soil Biology and Biochemistry*, 29 (1997), pp. 307–8.

25 Nico K. Michiels, A. Hohner and I. C. Vorndran, 'Precopulatory Mate Assessment in Relation to Body Size in the Earthworm *Lumbricus terrestris:* Avoidance of Dangerous Liaisons?', *Behavioural Ecology*, XII/5 (2001), pp. 612–18.

26 Visa Nuutinen and K. R. Butt, 'The Mating Behaviour of the Earthworm *Lumbricus terrestris* L. (Oligochaeta: Lumbricidae)', *Journal of Zoology* [London], 242 (1997), pp. 783–98.

27 Joris M. Koene, T. Pförtner and N. K. Michiels, 'Piercing the Partner's Skin Influences Sperm Uptake in the Earthworm *Lumbricus terrestris*', *Behavioural Ecology and Sociobiology*, 59 (2005), pp. 43–9.

28 Kevin R. Butt and V. Nuutinen, 'Reproduction of the Earthworm *Lumbricus terrestris* Linne After the First Mating', *Canadian Journal of Zoology*, LXXVI/1 (1998), pp. 104–9.

29 Gilbert White, *The Natural History of Selborne* (London, 1789), Letter XXXV, pp. 196–7.

30 Niki Grigoropoulou, K. R. Butt and C. N. Lowe, 'Interactions of Juvenile *Lumbricus terrestris* with Adults and Their Burrow Systems in a Two-Dimensional Microcosm', *Pesquisa Agropecuaria Brasileira*, 44 (2009), pp. 964–8.

31 Kevin R. Butt, 'Depth of Cocoon Deposition by Three Earthworm Species in Mesocosms', *European Journal of Soil Biology*, XXXVIII/2 (2002), pp. 151–3.

32 Nuutinen and Butt, 'Worms from the Cold', pp. 1580–82.

33 Kevin R. Butt, 'The Effects of Temperature on the Intensive Production of *Lumbricus terrestris* L. (Oligochaeta: Lumbricidae)', *Pedobiologia*, 35 (1991), pp. 257–64.

34 Kevin R. Butt and C. N. Lowe, 'Presence of Earthworm Species Within and Beneath *Lumbricus terrestris* (L.) Middens', *European Journal of Soil Biology*, XLIII (2007), pp. s57–s60.

35 Kevin R. Butt, J. Frederickson and R. M. Morris, 'The Life Cycle of the Earthworm *Lumbricus terrestris* L. (Oligochaeta: Lumbricidae) in Culture', *European Journal of Soil Biology*, 30 (1994), pp. 49–54.

36 Janice G. Mather and O. Christensen, 'Surface Migration of Earthworms in Grassland', *Pedobiologia*, 36 (1992), pp. 51–7.

37 Visa Nuutinen, 'The Meek Shall Inherit the Burrow: Feedback in Earthworm Soil Modification', in *Biology of Earthworms*, ed. A. Karaca (Heidelberg, 2011), pp. 123–40.

38 Grigoropoulou, Butt and Lowe, 'Interactions of Juvenile *Lumbricus terrestris*', pp. 964–8.

39 Kevin R. Butt and N. Grigoropoulou, 'Basic Research Tools for Earthworm Ecology', *Applied and Environmental Environmental Soil Science*, 29 November 2009, www.hindawi.com.

40 Niki Grigoropoulou and K. R. Butt, 'Field Investigations of *Lumbricus terrestris* Spatial Distribution and Dispersal through Monitoring of Manipulated, Enclosed Plots', *Soil Biology and Biochemistry*, XLIII/1 (2010), pp. 40–47.

41 Visa Nuutinen and K. R. Butt, 'Earthworm Dispersal of Plant Litter Across the Surface of Agricultural Soils', *Ecology*, C/7 (2019), E02669.

42 Kevin R. Butt, 'Food Quality Affects Production of *Lumbricus terrestris* (L.) Under Controlled Environmental Conditions', *Soil Biology and Biochemistry*, XLIII/10 (2011), pp. 2169–75.

43 Kevin R. Butt et al., 'An Oasis of Fertility on a Barren Island: Earthworms at Papadil, Isle of Rum', *Glasgow Naturalist*, 26 (2016), pp. 13–20.

44 Joshua Steckley, 'Cash Cropping Worms: How the *Lumbricus terrestris* Bait Worm Market Operates in Ontario, Canada', *Geoderma*, 343 (2020), no. 114128.

45 Paul F. Hendrix et al., 'Pandora's Box Contained Bait: The Global Problem of Introduced Earthworms', *Annual Reviews in Ecology, Evolution, and Systematics*, XXXIX (2008), pp. 593–613.

46 Nico Eisenhauer et al., 'Exotic Ecosystem Engineers Change the Emergence of Plants from the Seed Bank of a Deciduous Forest', *Ecosystems*, 12 (2009), pp. 1008–16.

47 Victor V. Pop and A. A. Pop, 'Lumbricid Earthworm Invasion in the Carpathian Mountains and Some Other Sites in Romania', *Biological Invasions*, 8 (2006), pp. 1219–22.

6 THE FUTURE OF WORMS AND HUMANS

1 These papers were published at www.scopus.com.
2 Lynette R. Potvin and E. A. Lilleskov, 'Introduced Earthworm Species Exhibited Unique Patterns of Seasonal Activity and Vertical Distribution, and *Lumbricus terrestris* Burrows Remained Usable for at Least 7 Years in Hardwood and Pine Stands', *Biology and Fertility of Soils*, 53 (2017), pp. 187–98.
3 British Trust for Ornithology (BTO), 'Bird Ringing Scheme', www.bto.org, accessed 4 September 2021.
4 Kevin R. Butt, M.J.I. Briones and C. N. Lowe, 'Is Tagging with Visual Implant Elastomer a Reliable Technique for Marking Earthworms?', *Pesquisa Agropecuaria Brasileira*, 44 (2009), pp. 969–74.
5 Niki Grigoropoulou and K. R. Butt, 'Field Investigations of *Lumbricus terrestris* Spatial Distribution and Dispersal through Monitoring of Manipulated, Enclosed Plots', *Soil Biology and Biochemistry*, XLI/1 (2010), pp. 40–47.
6 Chris N. Lowe, K. R. Butt and K. Y.-M. Cheynier, 'Assessment of Avoidance Behaviour by Earthworms (*Lumbricus rubellus* and *Octolasion cyaneum*) in Linear Pollution Gradients', *Ecotoxicology and Environmental Safety*, 124 (2016), pp. 324–8.
7 International Organization for Standardization, *Soil Quality – Avoidance Test for Determining the Quality of Soils and Effects of Chemicals on Behaviour, Iso 17512-1 – Part 1: Test with Earthworms (Eisenia Fetida and Eisenia Andrei)* (Geneva, 2008).
8 Sylvain Bart et al., '*Aporrectodea caliginosa*, a Relevant Earthworm Species for a Posteriori Pesticide Risk Assessment: Current Knowledge and Recommendations for Culture and Experimental Design', *Environmental Science and Pollution Research*, XXV/34 (2018), pp. 33867–81.
9 Valery E. Forbes et al., 'Mechanistic Effect Modeling of Earthworms in the Context of Pesticide Risk Assessment: Synthesis

of the FORESEE Workshop', *Integrated Environmental Assessment and Management*, XVII/2 (2020), pp. 352–63.

10 Pia Euteneuer et al., 'Earthworms Affect Decomposition of Soil-Borne Plant Pathogen *Sclerotinia sclerotiorum* in a Cover Crop Field Experiment', *Applied Soil Ecology*, CXXXVIII (2019), pp. 88–93.

11 Stefan Scheu, A. Theenhaus and T. Hefin Jones, 'Links Between the Detritivore and the Herbivore System: Effects of Earthworms and Collembola on Plant Growth and Aphid Development', *Oecologia*, CXIX/4 (1999), pp. 541–51.

12 Visa Nuutinen, K. R. Butt and L. Jauhiainen, 'Field Margins and Management Affect Settlement and Spread of an Introduced Dew-Worm (*Lumbricus terrestris* L.) Population', *Pedobiologia*, 54 (2011), pp. 167–72.

13 Kevin R. Butt, J. Frederickson and R. M. Morris, 'An Earthworm Cultivation and Soil-Inoculation Technique for Land Restoration', *Ecological Engineering*, IV/1 (1995), pp. 1–9.

14 Visa Nuutinen et al., 'Soil Faunal and Structural Responses to the Settlement of a Semi-Sedentary Earthworm *Lumbricus terrestris* in an Arable Clay Field', *Soil Biology and Biochemistry*, CXV (2017), pp. 285–96.

15 Kevin R. Butt and B. Williams, 'Vermiculture and Vermicomposting in the United Kingdom', in *Vermiculture Technology: Earthworms, Organic Wastes and Environmental Management*, ed. C. A. Edwards, N. Q. Aracon and R. Sherman (Boca Raton, FL, 2011), pp. 423–35.

16 Antonio Cortés et al., 'Unraveling the Environmental Impacts of Bioactive Compounds and Organic Amendment from Grape Marc', *Journal of Environmental Management*, CCLXXII (2020), no. 111066.

17 Anthony Bradshaw, 'Restoration of Mined Lands – Using Natural Processes', *Ecological Engineering*, VIII/4 (1997), pp. 255–69.

18 Kevin R. Butt and S. M. Quigg, 'Earthworm Community Development in Soils of a Reclaimed Steelworks', *Pedosphere*, XXXI/3 (2021), pp. 384–90.

19 Vincent Ducasse et al., 'Can *Lumbricus terrestris* Be Released in Forest Soils Degraded by Compaction? Preliminary Results from

Laboratory and Field Experiments', *Applied Soil Ecology*, CLXVIII (2021), no. 104131.

20 Kevin R. Butt, 'Earthworms in Soil Restoration: Lessons Learned from UK Case Studies of Land Reclamation', *Restoration Ecology*, XVI/4 (2008), pp. 637–41.

21 Francis Ashwood et al., 'Effects of Composted Green Waste on Soil Quality and Tree Growth on a Reclaimed Landfill Site', *European Journal of Soil Biology*, LXXXVII (2018), pp. 46–52.

22 Juliane Filser et al., 'Soil Fauna: Key to New Carbon Models', *Soil*, 2 (2016), pp. 565–82.

23 Jim Frederickson and G. Howell, 'Large-Scale Vermicomposting: Emission of Nitrous Oxide and Effects of Temperature on Earthworm Populations', *Pedobiologia*, 47 (2003), pp. 724–30.

24 Clive A. Edwards, 'Can Earthworms Harm the Planet?', *Biocycle*, XLIX/12 (2008), pp. 53–4.

25 Ingrid M. Lubbers et al., 'Earthworms Can Increase Nitrous Oxide Emissions from Managed Grassland: A Field Study', *Agriculture, Ecosystem and Environment*, CLXXIV (2013), pp. 40–48.

26 Helen R. P. Phillips et al., 'Global Distribution of Earthworm Diversity', *Science*, 316 (2019), pp. 480–85.

27 Adriana A. Pop, M. Wink and V. V. Pop, 'Use of 18s, 16s Rdna and Cytochrome C Oxidase Sequences in Earthworm Taxonomy (Oligochaeta, Lumbricidae)', *Pedobiologia*, 47 (2003), pp. 428–33.

28 R. Andrew King, A. L. Tibble and W.O.C. Symondson, 'Opening a Can of Worms: Unprecedented Sympatric Cryptic Diversity Within British Lumbricid Earthworms', *Molecular Ecology*, XVII (2009), pp. 4684–98.

29 Samuel W. James et al., 'DNA Barcoding Reveals Cryptic Diversity in *Lumbricus terrestris* L., 1758 (Clitellata): Resurrection of *L. herculeus* (Savigny, 1826)', *PLOS ONE*, 12 (2018), p. E15629.

30 David Porco et al., 'Biological Invasions in Soil: DNA Barcoding as a Monitoring Tool in a Multiple Taxa Survey Targeting European Earthworms and Springtails in North America', *Biological Invasions*, XV/4 (2013), pp. 899–910.

31 Erin K. Cameron, E. M. Bayne and M. J. Clapperton, 'Human-Facilitated Invasion of Exotic Earthworms into Northern Boreal Forests', *Ecoscience*, XIV/4 (2007), pp. 482–90.

32 Friederike Biernert et al., 'Tracking Earthworm Communities from Soil DNA', *Molecular Ecology*, XXI/8 (2012), pp. 2017–30.

33 Justine Lejoly, S. Quideau and J. Laganière, 'Invasive Earthworms Affect Soil Morphological Features and Carbon Stocks in Boreal Forests', *Geoderma*, 404 (2020), no. 115262.

34 Royal Botanic Gardens, Kew, 'Millennium Seed Bank', www.kew.org, accessed 4 September 2021.

35 'The Frozen Ark: Saving the DNA and Viable Cells of the World's Endangered Species', *Frozen Ark*, www.frozenark.org, accessed 4 September 2021.

Bibliography

Appelhof, Mary, *Worms Eat My Garbage* (Kalamazoo, MI, 1997)

Bogitsh, Burton J., C. E. Carter and T. N. Oeltmann, *Human Parasitology* (Washington, DC, 2018)

Darwin, Charles, *The Formation of Vegetable Mould Through the Action of Worms with Observations on Their Habits* (London, 1881)

Edwards, Clive A., and Norman Q. Arancon, *Biology and Ecology of Earthworms,* 4th edn (New York, 2022)

Edwards, Clive A., Norman Q. Arancon and Rhonda L. Sherman, eds, *Vermiculture Technology: Earthworms, Organic Wastes, and Environmental Management* (Boca Raton, FL, 2011)

Edwards, Clive A., and P. J. Bohlen, *Biology and Ecology of Earthworms* (London, 1996)

Le Clerc, Daniel, *A Natural History of Worms Bred in the Bodies of Men and Other Animals* (London, 1721)

Schwartz, Janelle A., *Worm Work* (Minneapolis, MN, 2012)

Sherlock, Emma, *Key to the Earthworms of the UK and Ireland*, 2nd edn (Telford, 2018)

Sherman, Rhonda, *The Worm Farmer's Handbook* (White River Junction, VT, 2018)

Sims, Reginald W., and B. M. Gerard, *Earthworms* (Shrewsbury, 1999)

Stewart, Amy, *The Earth Moved* (Chapel Hill, NC, 2004)

Wilson, Robert A., *An Introduction to Parasitology* (London, 1967)

BIBLIOGRAPHY FOR CHILDREN

Hargreaves, Roger, and Adam Hargreaves, *Walter the Worm*
    (London, 2018)
Jennings, Terry, *Earthworms* (Oxford, 1988)
Larson, Gary, *There's a Hair in My Dirt! A Worm's Story* (New York, 1998)

Associations and Websites

Across the world there are numerous groups with either an interest in worms in general, or in earthworms specifically. The list below provides a few examples of these, but there are many more.

12th International Symposium on Earthworm Ecology, July 2022
https://isee12.symposium.inrae.fr

American Council on Science and Health
www.acsh.org

BBC Teach, 'Science KS1/KS2: How fallen leaves are broken down by worms, fungi and slime moulds',
www.bbc.co.uk

Caenorhabditis elegans
www.wormbase.org

Dominguez, Lara, 'Worms',
www.pinterest.co.uk

Earthworm Society of Britain
www.earthwormsoc.org.uk

International Worm-Based Sanitation Association
www.iwbsa.org

LeMieux, Julianna, 'Why Worms are a Cornerstone of Scientific Research', American Council on Science and Health, 10 March 2017, www.acsh.org

Sherman, Rhonda, 'Wormy FACTS and Interesting Tidbits', NC State University,
https://composting.ces.ncsu.edu

The Soil Association
www.soilassociation.org

UCLan Earthworm Research Group
www.uclan.ac.uk

The Urban Worm
www.theurbanworm.co.uk

'Worms in Humans'
www.nhs.uk

Acknowledgements

This book is dedicated to scientists and non-scientists from across the world that I have had the privilege to work with and hold discussions with over many years. These include work colleagues and PhD students who may not be mentioned by name, but will recognize the contributions they have made.

Photo Acknowledgements

The author and publishers wish to thank the organizations and individuals listed below for authorizing reproduction of their work.

Andrés Rymel Acosta-Galvis (Batrachia Foundation): p. 27 bottom; Alamy: pp. 22 (Science Photo Library), 39 top (Minden Pictures), 57 (Paul Biggins), 104 (Album), 107 (Retro AdArchives), 116 (Tibbut Archive); Wormitecture © Myka Baum: p. 118; Biodiversity Heritage Library: p. 6; Kevin R. Butt: pp. 8, 23 top and bottom, 25, 31, 39 bottom, 46, 47, 52, 53, 59, 61, 62, 66, 68, 73, 76, 78, 83, 85 left and right, 92, 93, 96, 109, 114, 119, 124, 125 top, 127, 129, 130, 133, 134, 136, 137, 138, 143 left and right, 148, 154, 155, 157, 160, 166 top, 168; Centers for Disease Control and Prevention (CDC): p. 20; Alec Dempster (Isabelle Barois, ISEE9, 2010): p. 121; Elsevier.com: p. 147; ESA/UK Space Agency: p. 19; Timothy A.M. Ewin and Virgil Tanassa: p. 42; Jim Frederickson: p. 159; Niki Grigoropoulou: p. 142; Interplay Entertainment Corp.: p. 110; iStockphoto: p. 82 (power-offorever); Christopher N. Lowe: pp. 35, 49, 166 bottom; Metropolitan Museum of Art, New York: p. 15 (The Elisha Whittelsey Collection, The Elisha Whittelsey Fund, 1959. Public Domain); The National Museum of Dentistry, Baltimore, Maryland, USA: p. 99; Visa Nuutinen: p. 144; Public Domain: pp. 12, 81, 113; Richard Revels, FRPS: pp. 26, 28, 56, 125 bottom; Riverford Organic Farmers: p. 60; Royal Society of London: p. 65; Science Museum Group Collection: p. 100; Shutterstock: p. 55 (Eastimages); Paul Sorrell (500px.com/psorrell): p. 43; Chris Theobald: p. 153; Wikimedia Commons: pp. 10 (Public Domain), 11 (Public Domain), 27 top (Wilkinson M., Sherratt E., Starace F., Gower D. J. (2013)/CC BY 2.5 Generic License), 29 (Fastily at English Wikipedia/CC BY-SA 3.0 Unported

Index